Routledge Revivals

The Incommensurability Thesis

Originally published in 1994, *The Incommensurability Thesis* is a critical study of the Incommensurability Thesis of Thomas Kuhn and Paul Feyerabend. The book examines the theory that different scientific theories may be incommensurable because of conceptual variance. The book presents a critique of the thesis and examines and discusses the arguments for the theory, acknowledging and debating the opposing views of other theorists. The book provides a comprehensive and detailed discussion of the incommensurability thesis.

The Incommensurability Thesis

by Howard Sankey

Routledge
Taylor & Francis Group

First published in 1994
by Avebury, Ashgate Publishing Limited

This edition first published in 2018 by Routledge
2 Park Square, Milton Park, Abingdon, Oxon, OX14 4RN
and by Routledge
52 Vanderbilt Avenue, New York, NY 10017

Routledge is an imprint of the Taylor & Francis Group, an informa business

A Library of Congress record exists under LCCN: 95115103

ISBN 13: 978-0-367-26186-3 (hbk)
ISBN 13: 978-0-429-29191-3 (ebk)

The Incommensurability Thesis

HOWARD SANKEY
Department of History and Philosophy of Science
University of Melbourne
Victoria, Australia

Avebury
Aldershot · Brookfield USA · Hong Kong · Singapore · Sydney

Published by
Avebury
Ashgate Publishing Limited
Gower House
Croft Road
Aldershot
Hants GU11 3HR
England

Ashgate Publishing Company
Old Post Road
Brookfield
Vermont 05036
USA

British Library Cataloguing in Publication Data

Sankey, Howard
Incommensurability Thesis. - (Philosophy of Science Series)
I. Title II. Series
501

ISBN 1 85628 631 2

Contents

Acknowledgements

This book originated as a PhD thesis at the University of Melbourne, to which I am grateful for financial and material support. I wish to thank my supervisor Henry Krips, as well as John Clendinnen, Homer Le Grand and Rod Home for their various contributions to the project.

Some of the material included here has previously appeared in articles of mine in the following journals: *Australasian Journal of Philosophy*, *British Journal for the Philosophy of Science*, *Journal of Philosophical Research*, and *Studies in History and Philosophy of Science*. Full publication details of these articles are given in the bibliography.

I acknowledge permission of the editors of the following journals for permission to quote material: *Australasian Journal of Philosophy*, *History and Theory*, and *Inquiry*.

Permission to reproduce copyright material was granted by the following publishers: Blackwell Publishers, Cambridge University Press, New Left Books, and the Philosophy of Science Association.

Quotations appearing on pages 24, 26-7, 49-50, and 185 from works by Carl Kordig, Saul Kripke and Thomas S. Kuhn are reprinted by permission of Kluwer Academic Publishers.

Quotations on pages 14, 116, 118, 122-23, 128 and 130-1 from the following works of Donald Davidson and Paul Feyerabend are reprinted here by permission of Oxford University Press:

Donald Davidson, *Inquiries into Truth and Interpretation*, Clarendon Press, Oxford, 1984,
Paul Feyerabend, 'Putnam on Incommensurability', *British Journal for the Philosophy of Science*, Vol. 38, 1987, 75-81.

I acknowledge permission of the University of California Press to quote specific passages from the following work:
Jarrett Leplin (ed.), *Scientific Realism*, © 1984 The Regents of the University of California.

I am grateful to the University of Chicago Press for allowing me to include specific quotations from two books by Thomas S. Kuhn:
Thomas S. Kuhn: *The Structure of Scientific Revolutions*, Univer-sity of Chicago Press, Chicago, © 1962, 1970 by the University of Chicago,
The Essential Tension, University of Chicago Press, Chicago, © 1977 by the University of Chicago.

1 Incommensurability

1.1 An overview

The topic of this book is the problem of the incommensurability of scientific theories. The problem has to do with the nature of the semantical relations between the languages employed by scientific theories. Broadly speaking, to say that a pair of theories is incommensurable is to say that the theories do not share a common language, or that the terms they employ do not have common meaning.

Incommensurability stems from semantic dependence of the vocabulary employed by a theory upon the theoretical context in which it occurs. Such dependence leads to semantic variance between theories. The languages of competing or successive theories in the same domain may differ with respect to the meaning, and even the reference, of their terms. Hence, it may prove impossible to translate expressions of one theory into the language of another. Thus, to say that a pair of theories is incommensurable is to say that the languages of such theories fail either in whole or in part to be intertranslatable.

Translation failure has two closely related apparent consequences for theory comparison. Since such theories are expressed in different languages, no statement from one can formally contradict a statement from another: incommensurable theories are not in logical conflict. And since the content of such theories is expressed in languages with no common meaning, their content does not overlap: incommensurable theories are incomparable for content.

As it is generally understood, the incommensurability thesis combines these three claims. It is the thesis that the languages of some scientific theories are, at least in part, mutually untranslatable, and consequently there are no logical relations between them and their content is incomparable. Though it is standard to conjoin the three claims in this manner, they are logically distinct and neither stand nor fall together.

The incommensurability thesis is due to Paul Feyerabend and Thomas Kuhn, who presented it independently in writings first published in 1962.[1] Feyerabend's view that certain pairs of theories are incommensurable served as part of his critique of the empiricist idea that earlier theories are reducible to the later theories which replace them. Kuhn incorporated the idea of incommensurability into his account of scientific change as revolutionary transition between paradigms. As will be seen later in this chapter, they did not have precisely the same thing in mind, and both their views have evolved under pressure of criticism.

In proposing the incommensurability thesis, Kuhn and Feyerabend were reacting against the empiricist philosophy of science which was then dominant. Though they attacked empiricist orthodoxy on a number of fronts, their rejection of the empiricist idea of an independently meaningful and theory-neutral observation language is of most relevance to incommensurability. While empiricists were prepared to admit variation of meaning at the level of theoretical terms,[2] they held the meaning of observational vocabulary to be independent of theory. The existence of such an observational language would effectively have guaranteed what the incommensurability thesis denied, viz. a common semantic ground for theory comparison. For if the empirical consequences of rival theories were expressed in the observation language, then the theories might be directly compared in a shared vocabulary.

As against this tenet of empiricism, Kuhn and Feyerabend argued that observation is not itself an independent source of meaning and that the meaning of observational vocabulary in fact depends on theory. So the incommensurability thesis arose historically as a rejection of the empiricist idea of an observation language shared by and capable of arbitrating between theories. This suggests another way to characterize the incommensurability thesis, viz. as the denial of the existence of a theory-neutral language in which the content of theories may be compared.

The incommensurability thesis has stirred controversy because of a number of unattractive consequences. It threatens to undermine our image of science as a rational and progressive enterprise. On more extreme construals, it suggests a view of science on which communication failure is rampant, theory choice utterly irrational, and scientific progress a myth. Without going to such alarmist lengths, however, several implications of the thesis may be enumerated which do warrant concern.

One has to do with the very idea that competing theories may properly constitute rivals. For if alternative theories in the same domain are incommensurable, any genuine conflict between them appears to be precluded. This stems from the absence of logical conflict between incommensurable theories due to their formulation in different, untranslatable languages. If no statement of one theory may contradict any statement of a theory with which it is incommensurable, then it follows that there is no point of disagreement between the theories. It thus becomes unclear why there should be any need to compare or indeed choose between incommensurable theories in the first place.[3]

Now, in the absence of genuine rivalry between a pair of theories, the rationale for seeking grounds to decide between them is obscure. But even were the attempt to make such a choice not misguided, it is unclear how the choice could be made on rational grounds. For if theories neither agree nor disagree, it is not as if one might be shown to be a better account of the same phenomena. There is simply no point at issue between them. Such theories cannot be compared by means of a detailed comparison of their consequences with respect to a shared body of evidence. And without a common language in which to formulate conflicting empirical predictions, no crucial test is available. Nor, in general, could any means of theory appraisal which depends on content comparison be applicable in the choice between them.[4]

Incomparability of content leads also to problems with scientific progress. If theories are incomparable for content, then they cannot be shown to advance towards the truth by virtue of an increase of cumulative truth-content. And if theories are so semantically variant that they are about quite different things, then they cannot converge upon the truth by means of a build-up of truths about the same things.

The problems of rivalry, content comparison, and progress are the most important issues raised by the incommensurability thesis. They represent a challenge to the rationalist seeking to understand theory-choice as informed by a critical appraisal of genuinely alternative theories. They are a challenge to the realist inclined to view theory-change as resulting in an increase of truth about the world.

The broad outlines of a solution to all three problems may be derived from what may fairly be described as the standard response to the incommensurability thesis. Briefly, the standard response has been to grant change of meaning while insisting that semantically variant theories may still refer to the same things.[5] Theories whose terms refer to the same things may give incompatible accounts of shared objects of reference even if what they say about those objects is not expressed in a shared vocabulary. Since statements unlike in meaning may contain terms which overlap referentially, such statements may be incompatible despite difference of meaning. This enables a sense to be recovered in which theories may genuinely conflict and in which their consequences may be compared for points of disagreement. And since reference may be sustained to a common domain of entities in the transition between theories, it restores the capability of the advance of science to yield a growing fund of truth about a fixed set of entities.

While this approach has enjoyed wide popularity among critics of the incommensurability thesis, it is not without difficulties. The fundamental problem facing it is the problem of developing an account of reference capable of sustaining common reference between radically divergent theories. For in order for the approach to succeed, reference must be sufficiently independent of theory to be shared across theories and to survive theory transitions. Familiar examples of non-synonymous co-referential expressions (e.g. 'renate'/'cordate') illustrate the relationship called for by the approach. But such examples do not themselves constitute evidence that non-synonymous theoretical terms have the same reference. Nor do they indicate how to determine whether such terms co-refer. Since a theory would appear to play a crucial role in determining which entities its terms refer to, the possibility of theories failing to have common or overlapping reference to shared entities must be taken seriously. Thus it is today an urgent question for the critic of the incommensurability thesis whether in fact reference is independent of theory in the requisite sense.

The problem of how to respond to the incommensurability thesis within the framework of the theory of reference constitutes the problem-situation with which this book is concerned. Given the basic point about reference of the standard response, the first task is to find a suitable account of reference. Much of the literature on the topic has been devoted to this task. Thus in Chapter Two I will discuss the problems which have led, within the present context, to a widespread rejection of the traditional description theory of reference in favour of a causal theory of reference.[6] It will be found there, however, that problems with reference change and with the reference of theoretical terms necessitate modification of the basic causal theory of reference. The problem of reference change has led causal theorists to recognize that the reference of a given term may be fixed in more than one way.[7] The problem of the reference of theoretical terms leads to the recognition that certain types of description have a substantial role in fixing reference.[8]

Advocates of the standard response who adopt a modified causal account of reference may regard the modifications as insignificant concessions to the incommensurability thesis. For it seems not to have been widely appreciated that much that has been said on behalf of incommensurability can be embraced within the framework of the modified causal theory. In particular, it will be argued in Chapter Three on the basis of such a modified causal theory that the languages of certain purportedly incommensurable theories are indeed mutually untranslatable. This puts the critic of the incommensurability thesis in the peculiar position of having to defend the idea of an untranslatable language. Thus in Chapter Four the notion of translation failure between the languages of theories will be defended against influential and far-reaching criticism which has been directed against the very idea of an untranslatable language.[9]

Not insignificant concessions regarding continuity and commonality of reference ensue from the modifications as well. According to a view sometimes advanced by Kuhn and Feyerabend, in the transition between incommensurable theories a radical discontinuity of reference takes place.[10] It will be shown in Chapter Five that the view that there is wholesale change of reference depends on a description theory of reference and is consequently to be rejected. However, given the requisite modifications of the causal theory of reference, it is not open to us to say that reference is altogether invariant in theory-change. Thus a more moderate reference change view, such as Kuhn

espouses in recent writings,[11] converges with the modified causal-theoretic approach.

Sometimes a more extreme view is associated with the incommensurability thesis. Kuhn and Feyerabend occasionally suggest that, beyond mere difference of reference, incommensurable theories refer to different "worlds". Instead of differential linguistic relations to a fixed and independent reality, this involves a metaphysical thesis of the dependence of reality upon theory. Since the standard approach aims to show referential overlap between theories, it is a minimal precondition of the applicability of that approach that putatively incommensurable theories not be enclosed within disjoint "worlds" of theory-dependent entities. Thus in Chapter Six it will be argued, by primarily exegetical means, that Kuhn and Feyerabend are not idealists who deny the existence of a theory-independent reality. In Chapter Seven a variety of weaker "constructivist" positions will be criticized.

On the whole, this book is intended as a defence of the standard response. While it will be accepted that theories vary with respect to meaning and reference, it will be argued throughout that their content is comparable. Because this requires strict focus on issues germane to the referential relations between theories, no attempt will be made to discuss the broader issues of progress towards truth or rational theory-choice in any detail.

The remainder of this chapter is devoted to exposition of Feyerabend's and Kuhn's views on incommensurability. Feyerabend will be discussed in section 1.2, and Kuhn in 1.3.

1.2 Feyerabend

In this section Feyerabend's overall line of argument for incommensurability will be discussed. Feyerabend developed his view in a series of papers, most of which were originally published in the early and mid 1960's. Criticism of the view at first led him to introduce certain minor modifications. His later treatment of the issue has amounted at most to clarifications and extensions, which do not substantially affect the content of the idea. (Several of the relevant papers have been collected in Feyerabend (1981a) and will be referred to here as they appear in that volume.)

In brief, Feyerabend's idea of incommensurability arose as a result of his rejection of the empiricists' neutral observation

language. In his view, the meaning of observational terms is determined by the theory in which they occur. Since the basic ontology and conceptual apparatus of theories may differ, the terms defined within one theory may be indefinable in the context of another. Such theories cannot share any common statements. Hence, for Feyerabend, they are incommensurable in the sense that there is an "absence of deductive relations" between them (1978, p. 68), and they do "not possess any comparable consequences, observational or otherwise" (1981d, p. 93).

Initially, in his (1981c), Feyerabend attacked the idea that the meaning of observational language is determined by observation uncontaminated by theory. He argued that neither the pragmatic conditions in which such language is employed nor the phenomenological experience which accompanies observation determines observational meaning. Against what he termed the "principle of pragmatic meaning", he noted that regularity of linguistic performance in observational contexts does not determine meaning: "however well behaved and useful a human observer may be, the fact that in certain situations he (consistently) produces a certain noise, does not allow us to infer what this noise means" (1981c, p. 22). And against the "principle of phenomenological meaning", he argued that the immediate experience associated with the use of observation sentences does not determine meaning, but at most constitutes the cause of utterance. He concluded this initial stage of his attack with the suggestion that: "the interpretation of an observation language is determined by the theories which we use to explain what we observe, and it changes as soon as those theories change" (1981c, p. 31).

Feyerabend later broadened his attack on empiricism into an assault on the reductionist account of inter-theory relations.[12] According to reductionism, an earlier theory is either reduced to, or explained by, a later theory by means of logical derivation. That is, the reduced theory must be deductively entailed by the reducing theory. On this general model of theory replacement, later theories in a domain are comprehensive theories which subsume or contain the older theories already in the domain.[13]

On Feyerabend's analysis, this account of the relations between theories involves two key assumptions. Given that the reduced theory is to be deducible from the reducing theory, it assumes a "consistency condition": "only such theories are ... admissible in a given domain which either contain the theories already used in this domain, or which are at least consistent with them inside the domain" (1965, p. 164).[14] And given that a univocal vocabulary is necessary if the deduction is to be valid, it assumes a "condition of meaning invariance": "meanings will have to be invariant with respect to scientific progress; that is, all future theories will have to be framed in such a manner that their use in explanation does not affect what is said by the theories, or factual reports to be explained" (1965, p. 164). Quite apart from being entailed by the first condition, the condition of meaning invariance receives additional support from the empiricist assumption of a theory-neutral observation language.

Feyerabend's most fundamental objection to these two reductionist assumptions is that the condition of meaning invariance is violated in certain major changes of theory. As he puts it in the opening of his (1981d):

> What happens ... when a transition is made from a theory T' to a wider theory T (which ... is capable of covering all the phenomena that have been covered by T') is something much more radical than incorporation of the unchanged theory T' (unchanged, that is, with respect to the meanings of its main descriptive terms as well as to the meanings of the terms of its observation language) into the context of T. What does happen is, rather, a replacement of the ontology (and perhaps even of the formalism) of T' by the ontology (and the formalism) of T, and a corresponding change of the meanings of the descriptive elements of the formalism of T' (provided these elements and this formalism are still used). This replacement affects not only the theoretical terms of T' but also at least some of the observational terms which occurred in its test statements. (1981d, pp. 44-5)

The claim that in certain cases there is actually a change of meaning in the transition between theories implies that in such cases the older theories cannot be logically derived from the later ones which replace them. Feyerabend argues for the meaning variance of theoretical terms by considering crucial differences in the way basic concepts are defined within a number of opposing theories. He employs his criticism of the idea of an independent observation language to support the claim that such meaning variance extends to observational language. Feyerabend's view that meaning varies with change of theory suggests a view on which the meaning of terms employed in a theory is determined by the theoretical context in which they occur and varies with change of context. Indeed, Feyerabend explicitly endorses such a contextual account:

> the meaning of every term we use depends upon the theoretical context in which it occurs. Words do not "mean" something in isolation; they obtain their meanings by being part of a theoretical system. Hence if we consider two contexts with basic principles that either contradict each other or lead to inconsistent consequences in certain domains, it is to be expected that some terms of the first context will not occur in the second with exactly the same meaning. (1965, p. 180)

While this is certainly an explicit contextualism, Feyerabend's view of the theory-dependence of the meaning of observational terms is not simply that their meaning is determined by the context in which they occur.

This is because Feyerabend's contextual account has a close connection with his realism about theories. In claiming that the meaning of observational terms depends upon the theory in which they are employed, Feyerabend is advocating a "realistic interpretation of scientific theories" on which theories give meaning to their observational terms (1981d, pp. 51-3). According to Feyerabend,

> a realist ... wants to give a unified account, both of observable and unobservable matters, and he will use the most abstract terms of whatever theory he is contemplating for that purpose. He will use such terms in order either to give meaning to observation sentences or else to replace their customary interpretation. (1975, p. 279)

Thus, for Feyerabend, the meaning of observational terms does not depend on theory simply by virtue of context-dependence, but rather because realistically interpreted theories bestow meaning on the observational terms they employ.

It is not immediately clear, however, why realism should lead to the theory-dependence of observational term meaning. For if the issue of realism versus instrumentalism is taken to be whether theoretical expressions are to be construed as genuinely referential expressions, realism does not as such appear committed to theory-dependence. It is not inconsistent for a realist to construe theoretical terms as putatively referring expressions while denying that the meaning of observational terms is theory-dependent.

Here Feyerabend's idea seems to be that the meaning of observational terms is determined by theory precisely because theories do purport to describe reality, and because the ontology of such a theory has implications about the nature of observed entities. That is, given that meaning comes neither from experience nor application conditions, the meaning of an observational term, as used in a theory, depends upon the way in which the theory describes the entities to which such a term refers. So, while the meaning of observational terms does depend on theoretical context according to Feyerabend, it does so in the sense that it depends on the account provided by a given theory of the observable entities in its domain.[15]

There is a second sense in which Feyerabend's view is not mere contextualism. The thesis that meaning depends on theoretical context may initially suggest that the meaning of a term depends on its entire theoretical context, so that the whole theory is somehow constitutive of, or relevant to, the term's meaning.[16] Yet even in the contextualist remark quoted above, Feyerabend is clear that it is not theory considered as a whole, but a particular part of a theory ("basic principles"), which relevantly affects meaning. That is, the meaning of theoretical terms depends upon their connection with certain fundamental theoretical laws or postulates. This will emerge more clearly in discussing incommensurability, to which I now turn.

Feyerabend's denial of the reductionist thesis that later theories typically subsume earlier theories is encapsulated in his claim that certain pairs of theories are in fact incommensurable. As a first approximation, this amounts to the claim that the logical relations required for deductive subsumption fail to obtain.

Feyerabend introduces the concept of incommensurability in the course of an argument that the impetus theory is not reducible to Newtonian mechanics.[17] He considers a version of the law of inertia stated in terms of impetus and argues that it cannot be reduced to Newtonian mechanics because the concept of impetus cannot be appropriately related to Newtonian concepts. According to Feyerabend, the notion of impetus depends upon the Aristotelian principle that all motion is the result of the continuous action of some force (1981d, p. 62). He says that impetus was conceived as "an inner moving force" (1981d, p. 63) which was "the force responsible for the movement of the object that has ceased to be in direct contact, by push, or by pull, with the material mover" (p. 65). He then argues that "the 'inertial law' [ ] of the impetus theory is incommensurable with Newtonian physics in the sense that the main concept of the former, the concept of impetus, can neither be defined on the basis of the primitive descriptive terms of the latter, nor related to them via a correct empirical statement" (1981d, p. 76).

The reason 'impetus' cannot be defined in Newtonian terms is that the concept of impetus presupposes that sustained motion requires a cause. Since within Newtonian mechanics inertial motion is considered not to be under the influence of any forces, the concept of impetus depends on a principle which is incompatible with basic Newtonian assumptions. (Incidentally, this shows how specific parts of theories rather than the theoretical context as a whole influence the meaning of theoretical terms.)

The immediate consequence of incommensurability is that an earlier theory cannot be reduced by deductive subsumption to a later theory. Since the terms used to express the theories differ in meaning, no statement belonging to one may be derived from the other. Because the consequence classes of such theories have no elements in common Feyerabend says such theories are "deductively disjoint" (1978, p. 67) and that a comparative "judgment involving a comparison of content classes is ... impossible" (1981h, p. 153).

To further clarify the concept of incommensurability I will now discuss several complexities to which it gives rise. The first set of issues centers upon the question of how incommensurable theories may be about the same domain, and in what sense they may conflict. After discussing these issues I will consider Feyerabend's attempt to specify the sort of theory change which leads to incommensurability.

The first problem is that allegedly incommensurable theories constitute alternative accounts of the same domain. It is unclear how such theories may conflict with respect to the same domain and yet be logically unrelated. To put the point simply: in what sense can incommensurable theories be alternative accounts of the same thing if nothing asserted by one is denied by the other?

Feyerabend appears to hold that the problem can be removed, or at least minimized, by limiting incommensurability to general theories. He has frequently stressed that incommensurability is restricted to "comprehensive" or "non-instantial" theories (1981d, p. 44), which, unlike "all ravens are black", are "applicable to at least some aspects of everything there is" (1981f, p. 105). And he notes that:

> To circumvent the difficulty that arises when we want to say that incommensurable theories 'speak about the same things' I restricted the discussion to non-instantial theories... (1978, p. 68, fn. 118)

The reason for the restriction appears to be that comprehensive theories share no common observation language, whereas low-level theories may be compared with respect to an observation language provided there is a "background theory of greater generality that provides a stable meaning for observation sentences" (1965, p. 214). But restricting incommensurability to comprehensive theories does not avoid the problem. For if such theories are about everything in a common domain, then they would surely seem to be about at least some of the same things.

Feyerabend's main attempt to deal with the difficulty is his account of the comparison of incommensurable theories by crucial experiment. The fact that theories are subject to crucial test is problematic for his view: since incommensurable theories share no common statement there can be no prediction asserted by one and denied by the other. Feyerabend's account is based upon a "pragmatic theory of observation" according to which "observational statements are distinguished from other statements not by their meaning, but by the circumstances of their production" (1965, p. 212). Feyerabend distinguishes between an uninterpreted sentence and the statement expressed by the sentence under a given interpretation, so that the same sentence may express different statements. This account allows that observation sentences may continue to be applied in the same pragmatic conditions even though the meaning of such sentences varies with theoretical context (1965, p. 198). So while incommensurable theories do not share any common observation statement,

> there is still human experience as an actually existing process, and it still causes the observer to carry out certain actions, for example, to utter sentences of a certain kind... This is the only way in which experience judges a general cosmological point of view. Such a point of view is not removed because its observation statements say that there must be certain experiences that then do not occur... It is removed if it produces observation sentences when observers produce the negation of these sentences. It is therefore still judged by the predictions it makes. (1965, pp. 214-5)

Thus an observation sentence produced in response to an observation, and given a different meaning by two incommensurable theories, may nevertheless constitute the report of a crucial test which confirms one and refutes the other.[18]

The pragmatic account of observation explains how incommensurable theories may be applied to the same empirical domain, and be submitted to crucial test by the same experimental procedure. This means incommensurable theories may be genuinely rival accounts of a common domain if there are experimental procedures such that sentences produced in response to their outcome confirm one theory while disconfirming the other.

However, this is clearly a minimal sense of rivalry, since it is impossible for incommensurable theories to contradict each other. Yet the relation of incommensurability itself seems to involve a more robust relation of rivalry. For such theories have radically divergent ontological commitments, and indeed postulate the existence of quite distinct kinds of entities with respect to the same domain.[19]

Indeed, Feyerabend appears to take the relationship between the conceptual apparatus of incommensurable theories to be constitutive of a relation of rivalry in its own right. Originally, he even took this relation to involve inconsistency. Thus in his (1981d) he claims repeatedly that the concepts of incommensurable theories cannot be interdefined because "principles" or "laws" of the theories are inconsistent: e.g. "derivation (of T' from T) will ... be impossible if T' is part of a theoretical context whose rules of usage involve laws inconsistent with T" (1981d, p. 77). And in his (1965) he defined incommensurability in terms of inconsistency: "Two theories will be called incommensurable when the meanings of their main descriptive terms depend on mutually inconsistent principles" (1965, p. 227, fn. 19). If this were the case, then, notwithstanding the fact that the consequence classes of incommensurable theories have no intersection, such theories would still conflict in the sense that their basic principles are mutually contradictory.

However, in response to criticism, Feyerabend retreated from even this limited appeal to inconsistency. Shapere, for example, pointed out that: "sentences which do not have any common meaning can neither contradict, nor not contradict, one another".[20] He also challenged Feyerabend's definition of incommensurability in terms of the inconsistency of principles:

In what language are these "principles" themselves formulated? Presumably ... in order for them to be inconsistent with one another, they must be formulated, or at least formulable, in a common language. (1984b, p. 99, fn. 63)

Feyerabend accepted the point that, not sharing common statements, incommensurable theories are not mutually contradictory.[21] So any conflict inherent in the relationship between the concepts of incommensurable theories cannot be due to the inconsistency of the principles on the basis of which the concepts are defined.

In subsequent treatment of the issue, Feyerabend does not take the relation between incommensurable conceptual systems to involve inconsistency. Yet he continues to treat the relation as constitutive of some sort of disagreement. For he has repeatedly stressed that incommensurability is not simply a matter of difference of meaning between theories:

> mere difference of concepts does not suffice to make theories incommensurable in my sense. The situation must be rigged in such a way that the conditions of concept formation in one theory forbid the formation of the basic concepts of the other... (1978, p. 68, fn. 118)

Compare this with the following comment from one of his most recent papers:

> First, incommensurability as understood by me is a rare event. It occurs only when the conditions of meaningfulness for the descriptive terms of one language (theory, point of view) do not permit the use of the descriptive terms of another language (theory, point of view); mere difference of meaning does not yet lead to incommensurability in my sense. Secondly, incommensurable languages (theories, points of view) are not completely disconnected — there exists a subtle and interesting relation between their conditions of meaningfulness. (1987, p. 81)

If there were no connection at all between the concepts of incommensurable theories, then the relationship between the concepts could not itself be a source of rivalry. What Feyerabend has in mind instead is a relation of incompatibility between the principles which define the concepts of incommensurable theories.

Such incompatibility cannot reduce to the formal inconsistency of the statements of basic theoretical principles without revoking the above concession to Shapere.[22] In saying that the conditions of concept formation or meaningfulness of one theory forbid the formation of the concepts or the use of the terms of another theory, Feyerabend appears to be offering a new description of the relationship he had formerly described as the inconsistency of principles. For the reason that concepts of a theory cannot be defined within an incommensurable theory is that the principles needed to define them are rejected by the latter theory. Such basic principles constitute the conditions of concept formation (or meaningfulness) for the concepts (or terms) of the theory. And given that certain principles are rejected by a theory, no concept formed on the basis of such principles can be formed within the theory.

To see how the principles of such theories may be incompatible, due emphasis must be placed on the fact that incommensurable theories differ at the level of ontology.[23] As an example of a principle, we have already discussed the Aristotelian law that all sustained motion requires continuous causation.[24] The ontological import of such a principle is twofold: it leads to the postulation of entities of certain kinds; and it describes the behaviour of such entities. Thus the law that motion requires a cause leads to the postulation of a force, the impetus, which sustains the motion of projectile bodies. To postulate the existence of such an entity is to posit a force acting upon motions which are taken within the Newtonian context not to be subject to any cause. Since inertial motion is free from causal influence according to Newtonian physics, the assumption leading to the postulation of impetus is incompatible with Newton's physics.

The nature of the relation between theoretical principles is further elaborated in Feyerabend's discussion of the type of theory change which leads to incommensurability. Earlier, we took note of a charge commonly levelled against the idea that meaning depends on theoretical context: viz. it seems to follow from context-dependence that any alteration of theory results in a change of meaning. To meet the objection, Feyerabend gives a general specification of the extent of theoretical change necessary for incommensurability.

He reserves incommensurability for deep changes of theory involving change of ontology and conceptual framework. In the following passage Feyerabend distinguishes changes not affecting meaning from those leading to incommensurability:[25]

a diagnosis of stability of meaning involves two elements. First, reference is made to rules according to which objects or events are collected into classes. We may say that such rules determine concepts or kinds of objects. Secondly, it is found that the changes brought about by a new point of view occur within the extension of these classes and, therefore, leave the concepts unchanged. Conversely, we shall diagnose a change of meaning either if a new theory entails that all concepts of the preceding theory have zero extension or if it introduces rules which cannot be interpreted as attributing specific properties to objects within already existing classes, but which change the system of classes itself. (1981e, p. 98)

In short, alterations of theory within a stable system of concepts do not induce meaning change. But a change of theory on which displaced concepts fail to refer, or on which the system of classes is altered, involves change of meaning. More generally, if the kinds of entities to which a theory is committed are rejected by a later theory, then there is a change of meaning of the sort necessary for incommensurability.

It is important to note that the above passage suggests a more extreme thesis than mere conflict between the principles on which the concepts of theories are defined. It suggests that in the transition between conceptual systems, there is a major change of reference as well. For if all the displaced concepts have empty extensions and no terms of the new theory refer to any members of the old classes, then no terms of either theory refer to the same things. Thus there seems to be, beyond mere change of meaning, a radical discontinuity of reference between such theories.

This completes my exposition of Feyerabend's views on incommensurability. As I come to address specific issues within the remainder of the book, I will have occasion to consider in more detail various issues which the discussion in this section has only been able to touch upon.

1.3 Kuhn

While Feyerabend's views have changed little since originally being developed, Kuhn constantly restates his, and they have undergone continual revision. Exposition of Kuhn's position must trace his views through several historical stages of development. What follows is an attempt to provide a systematic interpretation of a diffuse and often confusing array of ideas. Emphasis will be placed on issues of direct relevance to this book, to the exclusion of other less germane topics.

Kuhn's treatment of incommensurability may be divided into early and late positions, separated by a transitional stage.[26] Originally, Kuhn's notion of incommensurability involved semantical, observational and methodological differences between the global theories which he called "paradigms". His initial discussion suggested that proponents of incommensurable theories are unable to communicate, and that there is no recourse to neutral experience or objective standards to adjudicate between theories. In subsequent efforts to clarify his position he restricted incommensurability to semantic differences, and assimilated it to Quinean indeterminacy of translation. During this intermediate stage Kuhn's treatment of the issues tended to be incomplete, often resulting in cursory discussion.[27] However, he has recently begun to develop his position in more refined form. His present view is that there is translation failure between a localized cluster of interdefined terms within the languages of theories.

Before proceeding with Kuhn, let me note three minor discrepancies between Kuhn and Feyerabend's accounts of incommensurability. First, although Kuhn initially included non-semantic factors, only later restricting it to the semantical, Feyerabend always restricted his use of the notion to the semantical sphere.[28] Second, Feyerabend takes incommensurability to affect the entirety of a theory's terms, whereas Kuhn tends to see it as localized. Third, for Feyerabend incommensurability is uncommon because it applies only to "non-instantial" theories, while for Kuhn it is more widespread since it occurs in scientific revolutions.

Incommensurability figures integrally in Kuhn's account of revolutionary scientific change in his (1970a). (Unless otherwise indicated, the page references for quotations from Kuhn's early position are to Kuhn (1970a).) According to Kuhn, scientific activity divides into periods of "normal science" punctuated at intervals by episodes of "revolution". Normal science is "research firmly based upon one or more past scientific achievements" (p. 10), and scientific revolutions are when "an older paradigm is replaced in whole or in part by an incompatible new one" (p. 92). The pivotal notion here is that of a "paradigm". Kuhn takes paradigms to be "universally recognized scientific achievements that for a time provide model problems and solutions to a community of practitioners" (p. viii); as such, they "provide models from which spring particular coherent traditions of scientific research" (p. 10). However, Kuhn also uses 'paradigm' in the broader sense of a global theoretical structure embracing the "network of commitments — conceptual, theoretical, instrumental, and methodological" (p. 42) of a normal research tradition.[29] Besides "tell[ing] us different things about the population of the universe and about that population's behaviour", paradigms "are the source of the methods, problem-field, and standards of solution accepted by any mature scientific community at any given time" (p. 103).

Revolutionary transition between paradigms is at the heart of Kuhn's account and is the point at which incommensurability enters. As it figures in Kuhn's account, incommensurability constitutes an impediment to choice of paradigm: "Just because it is a transition between incommensurables, the transition between competing paradigms cannot be made a step at a time, forced by logic and neutral experience" (p. 150). Because of incommensurability, the decision between rival paradigms does not admit of a neat resolution. Kuhn likens the process of choice to a "gestalt switch" (p. 150), and says "the transfer of allegiance from paradigm to paradigm is a conversion experience" (p. 151).

The influence of incommensurability is mainly apparent in paradigm debate: "the proponents of competing paradigms are always at least slightly at cross-purposes", and "fail to make complete contact with each other's viewpoints" (p. 148). The incommensurability which thus besets paradigm debate is due "collectively", Kuhn says, to the following three factors:

> [T]he proponents of competing paradigms will often disagree about the list of problems that any candidate for paradigm must resolve. Their standards or their definitions of science are not the same. (p. 148)

> Within the new paradigm, old terms, concepts, and experiments fall into new relationships one with the other. The inevitable result is ... a misunderstanding between the two competing schools... To make the transition to Einstein's universe, the whole conceptual web whose strands are space, time, matter, force, and so on, had to be shifted and laid down again on nature whole... Communication across the revolutionary divide is inevitably partial. (p. 149)

> In a sense that I am unable to explicate further, the proponents of competing paradigms practice their trades in different worlds ... practicing in different worlds, the two groups of scientists see different things when they look from the same point in the same direction. (p. 150)

Incommensurability thus emerges as a complex relation between paradigms consisting, at least, of standard variance, conceptual disparity, and theory-dependence of observation.

The thesis that there may be no appeal to neutral observation and that standards of theory appraisal are internal to paradigm suggests a relativistic view of the epistemic merits of paradigms.[30] For if, in the absence of independent means of evaluating paradigms, a paradigm is to be assessed by standards dictated by the paradigm itself, such appraisal is relative to acceptance of paradigm. Yet Kuhn has resisted the charge of relativism, maintaining instead that there are shared scientific values independent of paradigms.[31] However, he insists that such values fail to unambiguously determine choice of theory. This enables him to restate the problem of deciding between paradigms:

> There is no neutral algorithm of theory-choice, no systematic decision procedure which, properly applied, must lead each individual in the group to the same decision. (1970a, p. 200)

Since we are concerned here with incommensurability as a semantical issue, and since Kuhn later separates such methodological issues from incommensurability proper, I will not pursue the theme of standard variation any further. Instead, I will now focus upon the semantic and conceptual aspects of Kuhn's early account of incommensurability.

The second factor contributing to incommensurability involves change of conceptual apparatus: "to make the transition to Einstein's universe, the whole conceptual web whose strands are space, time, matter, force, and so on, had to be shifted and laid down again on nature whole" (p. 149). In a manner reminiscent of Feyerabend, Kuhn takes such conceptual change to prevent the laws of a displaced paradigm from being derived from the paradigm which replaces it.

Kuhn argues that the analogues of Newton's laws that follow from Einstein's physics as a special case are not identical with those laws. This is because the statements of Einsteinian versions of the laws employ relativistic concepts which "represent

Einsteinian space, time, and mass", and so differ in meaning from the statements which express Newton's laws:

> the physical referents of these Einsteinian concepts are by no means identical with those of the Newtonian concepts that bear the same name. (Newtonian mass is conserved; Einsteinian is convertible with energy. Only at low relative velocities may the two be measured in the same way, and even then they must not be conceived to be the same.) Unless we change the definitions of the variables in the [Einsteinian versions of the laws], the statements we have derived are not Newtonian ... the argument has [ ] not done what it purported to do. It has not, that is, shown Newton's Laws to be a limiting case of Einstein's. For in the passage to the limit it is not only the forms of the laws that have changed. Simultaneously we have had to alter the fundamental structural elements of which the universe to which they apply is composed. (pp. 101-2)

This passage reveals a fundamental convergence between Kuhn's and Feyerabend's notions of incommensurability. For, as with Feyerabend's original use of the notion, Kuhn's argument against the derivation of Newton's laws from Einstein's is directed against the reductionist account of theory replacement.

Indeed, since the failure of derivability is due to conceptual disparity between the theories, Kuhn's notion of incommensurability may even appear to coincide with Feyerabend's exactly.[32] The equivalence of their views is strongly suggested as well by the fact that Kuhn combines the claim of conceptual disparity with a rejection of the empiricists' neutral observation language.[33] For this suggests that with Kuhn, as with Feyerabend, incommensurability does not consist simply in difference of the basic concepts of theories: it also involves dependence of the meaning of observational terms upon the theory in which they occur. However, Kuhn later claimed only to have meant that part of the languages of incommensurable theories differ in meaning.[34] This attenuates the parallel between Kuhn's original notion of semantical incommensurability and Feyerabend's. For it suggests that the language used to report observations, while not being theory-neutral, is only in part semantically variant between theories.

While this implies that incommensurable paradigms are not altogether unrelated semantically, Kuhn is sometimes drawn toward a far stronger thesis. This is apparent from the third constitutive element of incommensurability: viz. that "proponents of competing paradigms practice their trades in different worlds" (p. 150). Kuhn's (1970a) contains numerous comments to the effect that "when paradigms change, the world itself changes with them" (p. 111), and "after a revolution scientists work in a different world" (p. 135). Although the image of "world-change" is usually qualified in some way, it suggests that the transition between incommensurable paradigms is a transition from the "world" of one paradigm to the "world" of another. Often, such remarks are meant only to emphasize the influence of conceptual framework on perception, as in this comment on the failure to derive Newton's laws from Einstein's:

> the transition from Newtonian to Einsteinian mechanics illustrates with particular clarity the scientific revolution as a displacement of the conceptual network through which scientists view the world. (p. 102)

At other times, Kuhn intends the difference to go beyond difference of perception:

> paradigm changes do cause scientists to see the world of their research-engagement differently. In so far as their only recourse to that world is through what they see and do, we may want to say that after a revolution scientists are responding to a different world. (p. 111)

> in the absence of some recourse to that hypothetical fixed nature that he "saw differently," the principle of economy will urge us to say that after discovering oxygen Lavoisier worked in a different world. (p. 118)

In such passages, Kuhn seems inclined to view the world independent of scientific belief and perception as dispensable.

Kuhn wishes to say that incommensurable paradigms present scientists with different "visual gestalts" of the same world (cf. pp. 111-2). And he insists that "though the world does not change with change of paradigm, the scientist afterward works in a different world" (p. 121). Yet his tendency to dispense with the world beyond the perceptual and epistemic states of the scientist strongly suggests that there is nothing over and above the "world" presented by the gestalt of a paradigm, or at least that the world in itself is of no relevance to science. The tension between admitting an independent reality and discarding it is never clearly resolved in Kuhn's original account, and has resulted in the widespread impression that his version of incommensurability involves some form of idealism.[35]

However, the "world-change" image may also be interpreted in a weaker sense as expressing a thesis about reference. It may be taken as the idea that there is a major difference in reference between paradigms. This interpretation is suggested by Kuhn's previously quoted discussion of Newtonian and Einsteinian concepts (pp. 101-2). In that passage Kuhn asserts that "the physical referents of these Einsteinian concepts are by no means identical with those of the Newtonian concepts that bear the same name". And he remarks that "Newtonian mass is conserved; Einsteinian is convertible with energy", which suggests that the terms for mass in the two theories do not have the same reference. In the light of such remarks, the "world-change" image may be taken to mean that in the transition between incommensurable paradigms there is a wholesale change in what is referred to. Thus, talk of the "world" of a theory may be construed as talk about the set of entities to whose existence the theory is committed and to which its terms purportedly refer.

In sum, not even the conceptual component of Kuhn's original diffuse notion of incommensurability admits of unified analysis. Paradigms which are incommensurable due to conceptual variance are not derivable from one another; in some sense, they may even be about different worlds; or perhaps they simply fail to have common reference. These disparate elements begin to coalesce during Kuhn's transitional phase, which we will now consider.

In subsequent development of his views, three general points emerge as basic to Kuhn's position. First, direct comparison of theories requires their formulation in a common language: "The point-by-point comparison of two successive theories demands a language into which at least the empirical consequences of both can be translated without loss or change" (1970b, p. 266). Second, no such common language is available: "There is no neutral language into which both of the theories as well as the relevant data may be translated for purposes of comparison" (1979, p. 416). Third, exact translation between the languages of theories is impossible: "translation of one theory into the language of another depends ... upon compromises ... whence incommensurability" (1976, p. 191). Thus, in clarifying incommensurability, the issue of translation failure between theories becomes the dominant theme.

Reflection on translation has led Kuhn to draw a connection between incommensurability and Quine's thesis of the indeterminacy of translation.[36] Quine's thesis, in brief, is that "manuals for translating one language into another can be set up in divergent ways, all compatible with the totality of speech dispositions, yet incompatible with one another" (1960, p. 27). The thesis stems from a behaviourist critique of meaning: Quine holds that verbal behaviour leaves meaning indeterminate; and he denies there are facts about meaning beyond what is evident in such behaviour. The key to the thesis is an indeterminacy in the reference of sortal predicates, as illustrated by Quine's imagined native word 'gavagai' (1960, p. 52). Quine argues that the reference of 'gavagai' is inscrutable: ostension does not determine whether it refers to rabbits, rabbit-stages, or undetached rabbit parts (1969, p. 30); while the translation of the native "individuative apparatus" needed for fine discrimination of reference is also indeterminate (1969, p. 33). Inscrutability of reference renders the translation of sentences containing such terms indeterminate.

At times Kuhn draws support from the indeterminacy thesis. In arguing that translation "always involves compromises", Kuhn cites Quine's discussion of indeterminacy as evidence that "it is today a deep and open question what a perfect translation would be and how nearly an actual translation can approach the ideal" (1970b, p. 268). He appeals to Quine's 'gavagai' example to indicate the epistemological difficulties of translating a language with different concepts:

> Quine points out that, though the linguist engaged in radical translation can readily discover that his native informant utters 'Gavagai' because he has seen a rabbit, it is more difficult to discover how 'Gavagai' should be translated... Evidence relevant to a choice among [ ] alternatives will emerge from further investigation, and the result will be a reasonable analytic hypothesis... But it will be only a hypothesis... [T]he result of any error may be later difficulties in communication; when it occurs, it will be far from clear whether the problem is with translation and, if so, where the root difficulty lies. (1970b, p. 268)

At a later stage, however, Kuhn seeks to distance his position from Quine's. In the following passage he explains how his views on reference and translation diverge from those of Quine:

> Unlike Quine, I do not believe that reference in natural or scientific languages is ultimately inscrutable, only that it is very difficult to discover and that one may never be absolutely certain one has succeeded. But identifying reference in a foreign language is not equivalent to producing a systematic translation manual for that language. Reference and translation are two problems, not one, and the two will not be resolved together. Translation always and necessarily involves imperfection and compromise; the best compromise for one purpose may not be the best for another; the able translator, moving through a single text, does not proceed fully systematically, but must repeatedly shift his choice of word and phrase, depending on which aspect of the original it seems most important to preserve. (1976, p. 191)

As opposed to Quine, Kuhn holds that while it may be determined what the terms of another language or theory refer to, they may prove not to be translatable in a faithful or uniform manner.

Kuhn's appeal to Quine is somewhat misleading, since it tends to suggest that incommensurability is a form of the indeterminacy of translation. For Quine, translation is indeterminate in the sense that there is no fact of the matter about how to translate from one language into another: indeterminacy means no sense can be made of correct translation. Kuhn's claim that translation involves compromise and imperfection runs counter to indeterminacy since it presupposes that, at least in principle, correct translation is possible: translation is only compromised if there is something to be right about.[37] As will become clear in the sequel, for Kuhn incommensurability implies failure of exact translation between theories: terms of one theory have meaning which cannot be expressed within the language of another theory. As such, the claim of incommensurability denies translation in a manner which is impossible if translation is indeterminate in Quine's sense.

Despite treating translation as the basic issue, Kuhn does not provide a detailed analysis of translation failure between theories during this transitional period. What little he does say amounts at most to a general indication of the cause and extent of such failure. Kuhn explains that translation is problematic, "whether between theories or languages", because "languages cut up the world in different ways" (1970b, p. 268). Theories employ different systems of "ontological categories" (1970b, p. 270) in order to classify the objects in their domain of application. In the transition between theories classificatory schemes change:

> One aspect of every revolution is, then, that some of the similarity relations change. Objects which were grouped in the same set before are grouped in different sets afterwards and vice versa. Think of the sun, moon, Mars, and earth before and after Copernicus; of free fall, pendular, and planetary motion before and after Galileo; or of salts, alloys, and a sulphur-iron filing mix before and after Dalton. Since most objects within even the altered sets continue to be grouped together, the names of the sets are generally preserved. (1970b, p. 275)

Such categorial change involves change in the meaning, and even the reference,[38] of the retained terms:

> In the transition from one theory to the next words change their meanings or conditions of applicability in subtle ways. Though most of the same signs are used before and after a revolution — e.g. force, mass, element, compound, cell — the ways in which some of them attach to nature has somehow changed. Successive theories are thus [ ] incommensurable. (1970b, p. 267)

Since it is only some of the "similarity-sets" that change, and only some terms "attach to nature" differently, the translation failure resulting from such conceptual change is of limited scope.

Apart from the claim that translation between theories involves compromise and imperfection, Kuhn does little at this stage to clarify the semantical aspects of such translation failure. On occasion Kuhn oversimplifies the issue by writing as if change in meaning of retained terms were in itself sufficient for untranslatability. In the preceding quotation, for example, Kuhn's inference from change of meaning to incommensurability is direct and without qualification. Elsewhere he claims that scientists who "perceive the same situation differently" while using common vocabulary "must be using words differently", and hence speak from "incommensurable viewpoints" (1970a, p. 200). Such a pattern of inference suggests that assigning different meanings to old terms is all that is required for incommensurability to occur.

But this makes the connection between change of meaning and incommensurability too direct. If incommensurability involves failure to translate from one theory into another, mere change in the meaning assigned to shared words does not in itself suffice for incommensurability. The point is simply that a vocabulary can undergo change of meaning without necessarily resulting in failure to translate. For one thing, such a change in the meaning of words can occur in a trivial manner: words may have their meanings switched around. A fixed stock of meanings may be reassigned to different terms of a given vocabulary without leading to translation failure between the alternative interpretations of the vocabulary.

Less trivially, single words with identical meanings are unnecessary for translation: translation need not be word-for-word. Even if there are terms in one language not matched by individual words the same in meaning in the other language, it may still be possible to translate them by combinations of terms, or phrases, of the other language. Hence a change in the meaning of some of the terms which are retained between theories need not lead to an inability to translate from the language of one theory into that of another. The general point is that what is needed for translation failure is something more than mere change of meaning. At the very least, Kuhn's claim of partial translation failure requires an inability on the part of some theory to define terms which are employed within another theory.[39]

A further source of unclarity is Kuhn's treatment of the relation between translation and comparison of content. As we noted earlier, Kuhn takes "point-by-point comparison" of theories to require formulation in a common language (1970b, p. 266). And he takes incommensurability to imply that theories are unable to be compared in such a manner:

> In applying the term 'incommensurability' to theories, I had intended only to insist that there was no common language within which both could be fully expressed and which could therefore be used in a point-by-point comparison between them. (1976, p. 191)

Yet Kuhn also denies that incommensurability is to be construed as incomparability:

> Most readers [ ] have supposed that when I spoke of theories as incommensurable, I meant that they could not be compared. But 'incommensurability' is a term borrowed from mathematics, and it there has no such implication. The hypotenuse of an isosceles right triangle is incommensurable

with its side, but the two can be compared to any required degree of precision. What is lacking is not comparability but a unit of length in terms of which both can be measured directly and exactly. (1976, p. 191)

This is puzzling, for it raises the question of how the content of theories inexpressible in a common language can be compared, if not in point-by-point manner.[40]

However, while denying comparison in a common language, Kuhn notes that "comparing theories ... demands only the identification of reference" (1976, p. 191), and that "systematic theory comparison requires determination of the referents of incommensurable terms" (1976, p. 198, fn. 11). Although he fails to elaborate, Kuhn is implicitly contrasting "point-by-point" comparison with comparison by means of reference. He does not explain what "point-by-point" comparison is, but he seems to be operating with a distinction between direct comparison of statements expressed in a common vocabulary and comparison of statements which differ in meaning via overlapping reference.

More specifically, two theories which share a common vocabulary invariant in meaning may diverge simply with respect to the truth-values they assign to a common set of statements. Such theories may be compared "point-by-point" in the sense that one theory asserts precisely the same statement that the other denies. By contrast, theories expressed in vocabulary which is variant with respect to meaning may still be compared by means of overlapping reference. Such theories do not assert or deny a common set of statements. But even if their statements do not have the same meaning, they may be compared if the constituent terms of their statements have the same reference. Such a comparison fails to be "point-by-point" because it does not consist in pairing a statement asserted by one theory with its denial drawn from another theory. It may also fail to be "point-by-point" in another sense: since not all terms of one theory need co-refer with terms of the other, not all statements of the theories may be brought into conflict by means of relations of co-reference.[41]

To conclude discussion of Kuhn's middle period, recall the disparate elements of his original position mentioned earlier. Kuhn's original conception involved failure of derivation, "world-change" and wholesale change of reference. The picture which emerges from this transitional phase combines these elements in more coherent fashion. It remains the case that the central statements of a theory are not entailed by a theory with which it is incommensurable. But given Kuhn's restriction of change of meaning and reference to only some of a theory's terms, it follows that incommensurable theories share a modicum of semantically invariant vocabulary. As a result, there is neither complete change of reference, nor does the world which theories are about change. Thus, Kuhn's "world-change" image may be interpreted as change in the basic "ontological categories" which different theories impose upon the world.

Incommensurability, as portrayed during Kuhn's middle period, involves partial translation failure between theories committed to different basic categories. Though such broad features of Kuhn's position subsequently remain unaltered, the details are refined in more recent work, especially his (1983). Kuhn's later position is characterized by a more nuanced account of translation failure and its connection with categorial change.

In his (1983) Kuhn outlines a notion of "local incommensurability" which he claims to have been his original idea.[42] In brief, local incommensurability consists in failure to translate between localized clusters of interdefined terms:

> The claim that two theories are incommensurable is [ ] the claim that there is no language, neutral or otherwise, into which both theories, conceived as sets of sentences, can be translated without residue or loss... Most of the terms common to the two theories function the same way in both; their meanings, whatever they may be, are preserved; their translation is simply homophonic. Only for a small subgroup of (usually interdefined) terms and for sentences containing them do problems of translatability arise. (1983, pp. 670-1)

So construed, incommensurability is a limited inability to translate from a local subgroup of terms of one theory into another local subgroup of terms of another theory. As such, language peripheral to the non-intertranslatable subgroups of terms constitutes semantic common ground between incommensurable theories. Hence, as Kuhn admits (1983, p. 671), at least part of the content of such theories may be directly compared.

Kuhn continues to link translation failure closely with change of classification, maintaining, as previously, that the membership classes of certain key categories are altered in the transition between incommensurable theories. Since the categories are interrelated, such changes are not isolated, but have a holistic effect:

> What characterizes revolutions is [ ] change in several of the taxonomic categories prerequisite to scientific descriptions and generalizations. That change, furthermore, is an adjustment not only of criteria relevant to categorization, but also of the way in which given objects and situations are distributed among pre-existing categories. Since such redistribution always involves more than one category and since those categories are interdefined, this sort of alteration is necessarily holistic. (1981, p. 25)

Kuhn explains, in his (1983, pp. 682-3), that languages and theories deploy sets of "taxonomic categories" constitutive of "taxonomic structures". In translating between them, it is necessary to preserve categories; and, because of the interconnection of categories, translatable languages must have the same taxonomic structure. Translation problems arise because "different languages (and theories) impose different structures on the world" (p. 682); for translation to succeed, "taxonomy must [ ] be preserved to provide both shared categories and shared relationships between them" (p. 683).

The holistic nature of category change is directly reflected in translation failure: the interconnection of categories is paralleled by the interdefinition of concepts. Kuhn illustrates this with examples,[43] arguing, for instance, that while much language used in phlogistic chemistry is subsequently retained, "a small group of terms remains for which the modern chemical vocabulary offers no equivalent" (1983, p. 675). The residual terms, which include 'phlogiston' and its cognates, as well as 'element' and 'principle', constitute an interdefined cluster not definable within later theory. While Kuhn grants that various applications of such terms may be specified in the language of modern theory, he denies that translation is possible:

> Among the phrases which describe how the referents of the term 'phlogiston' are picked out are a number that include other untranslatable terms like 'principle' and 'element'. Together with 'phlogiston', they constitute an interrelated or interdefined set that must be acquired together, as a whole, before any of them can be used, applied to natural phenomena. Only after they have been thus acquired can one recognize eighteenth-century chemistry for what it was, a discipline that differed from its twentieth-century successor not simply in what it had to say about individual substances

and processes but in the way it structured and parceled out a large part of the chemical world. (1983, p. 676)

Translation between such local complexes of terms fails because the meaning of such terms is determined in relation to other terms of the interdefined set. Terms which are defined within an integrated set of concepts cannot be translated in piecemeal fashion into an alternative complex in which the necessary conceptual relations do not obtain.[44]

The notion of a localized translation failure between interdefined sets of terms is the central feature of Kuhn's later account of incommensurability and the most significant refinement of his position. As we saw earlier, the thesis of local incommensurability was neither developed in detail nor clearly evident in Kuhn's original discussion of the issue. While the local thesis is suggested obliquely during his middle period, explicit development of the local version constitutes a further step in the overall process of moderation which Kuhn's account of incommensurability has undergone. Because the local version of the thesis confines untranslatability to localized clusters of terms, Kuhn's later position also represents a weaker account of incommensurability than Feyerabend's. Since Feyerabend takes the semantical differences constitutive of incommensurability to extend to the observation terms employed by such theories, on his view failure of intertranslation must apply to the entirety of the language they employ.

Notes

1. See Feyerabend (1981d) and Kuhn (1970a).
2. E.g. see the discussion of Carnap in English (1978); cf. Newton-Smith (1981, p. 152).
3. For a precis of various arguments from rivalry, see Kordig (1971, pp. 52-5).
4. This has led a number of writers to propose standards of comparison which do not depend on content comparison: see Laudan (1977, pp. 143-5) on problem-solving, Leplin (1979, p. 266) on formal and evaluative standards, and Moberg (1979, pp. 257-8) on internal consistency.
5. This standard response stems from Scheffler (1967) and Putnam (1975a). It will be discussed in depth in Chapter Two.

6. Putnam (1975a) and (1975b) are the most relevant original developments of the causal theory of reference in the present context.
7. See Devitt (1979, p. 41) on "multiple grounding" and Kitcher (1978) on the different ways in which the tokens of a term-type may be fixed.
8. See the discussion of theoretical term reference in Kroon (1985).
9. The objections are due to Davidson (1984) and Putnam (1981, pp. 114-5).
10. See particularly Feyerabend (1981e, p. 98) and Kuhn (1970a, p. 102).
11. E.g. see Kuhn (1979) and (1981).
12. See especially Feyerabend (1981d) and (1965).
13. The classic source for the reductionism to which Feyerabend was reacting is Nagel (1961). For a brief discussion of the view see Suppe (1977, pp. 53-6).
14. In his (1981d) Feyerabend speaks instead of a "principle of deducibility", according to which "explanation is achieved by deduction in the strict logical sense" (1981d, p. 46).
15. Cf. Feyerabend (1965, p. 170): "For example, we may change our ideas about the nature, or the ontological status (property, relation, object, process, etc.) of the color of a self-luminescent object without changing the methods used for ascertaining that color (looking, for example). Clearly, such a change is bound profoundly to influence the meanings of our observational terms."
16. This is exactly what Feyerabend's contextualist remark has suggested to critics. E.g. Shapere interprets the quoted passage as "suggesting that the slightest alteration of theoretical context alters the meaning of every term in that context" (1984b, p. 71).
17. Feyerabend (1981d, pp. 65-7). He applies the same pattern of argument to other cases, such as the case of Einsteinian and Newtonian concepts of mass; see especially (1965, pp. 168-9).
18. It is not even necessary for there to be a single common observation sentence produced in response to the test. The same experimental result may be described by theories using completely different terminology and still count in favour of one theory and against the other. See Feyerabend (1975, p. 282).

19. It will be argued later (5.2) that the pragmatic account of observation is committed at least to referential overlap at the observational level, so that a substantial rivalry relation obtains between such theories.
20. Quoted in Feyerabend (1981f, p. 115) from a letter of Shapere's. The relevant criticism is in Achinstein (1964, p. 499) and Shapere (1984, p. 73).
21. See Feyerabend (1981f, p. 115). After conceding Shapere's point, Feyerabend attempted to show how incommensurable theories may be incompatible without being mutually contradictory. The most definite suggestion is that their ontologies may have different structures: "simply compare two infinite sets of elements with respect to certain structural properties and inquire whether or not an isomorphism can be established" (1981f, p. 115). He also suggests several ways in which the pragmatic account of observation enables incompatibility to be shown by comparison of incommensurable theories with respect to empirical evidence (1981f, p. 116).
22. In his (1975) Feyerabend tries to avoid Shapere's objection to the inconsistency of principles as follows: "by a 'principle' I do not simply mean a statement ... but the grammatical habit corresponding to the statement" (1975, p. 270). Though the notion of a "grammatical habit" is obscure, the point is presumably that habits do not enter logical relations such as inconsistency. But even if it were not obscure, it would not explain the sense in which the principles are at variance with one another.
23. In arguing that incommensurable theories share no statements, Feyerabend stresses that: "we very often discover that entities we thought existed did, in fact, not exist. Realizing this, we must eliminate and replace the terms designating these entities from our factual descriptions" (1965, p. 170). See also (1975, p. 275).
24. Another example used by Feyerabend is the Newtonian principle that "properties of physical objects, such as shapes, masses, volumes, time intervals ... inhere in objects and change only as the result of a direct physical interference" (1975, p. 275).
25. Though the passage does not explicitly connect meaning change with incommensurability, when the distinction is applied to the spatio-temporal concepts of classical mechanics and general relativity, Feyerabend uses the

distinction to "diagnose" incommensurability (1981e, pp. 99-100).

26. The main body of Kuhn's (1970a) will be taken as the source of his early position. The transitional phase is represented by the 'Postscript' to his (1970a), his (1970b), (1976) and (1979). His later position is found in his (1981) and (1983).
27. Kuhn's first main attempts at clarification were published around 1970; see the 'Postscript' to his (1970a) and his (1970b). Over the next ten years he was somewhat reticent and his discussion of incommensurability was confined to brief remarks in his (1976) and (1979).
28. Feyerabend mentions this difference between his own notion of incommensurability and Kuhn's original use of the notion in his (1978, pp. 66-7).
29. The ambiguity of Kuhn's original use of 'paradigm' has been widely noted; see, for example, Shapere (1984a, p. 39) and Masterman (1970). Kuhn subsequently distinguished the paradigm as "constellation of beliefs, values, techniques" from the paradigm as "shared example", referring to them as 'disciplinary matrix' and 'exemplar' respectively; see the 'Postscript' to his (1970a) as well as his (1977a).
30. Kuhn's seeming denial of extra-paradigmatic criteria of theory-choice has appeared relativist and irrationalist to many commentators. See, for example, Scheffler (1967, pp. 74ff) and Shapere (1984a, p. 46).
31. Kuhn lists such cognitive values as accuracy, simplicity, fruitfulness, internal and external consistency; see his 'Postscript' (1970a, pp. 185, 199). He discusses the issues raised by differential weighting of values and variant application of the same value in his (1977b).
32. Shapere, for example, explicitly equates their views, see his (1984b, p. 83); the equation is implicit in Scheffler (1967, pp. 49-50).
33. For the rejection, see Kuhn (1970a, pp. 125-9). Kuhn's main argument seems to be as follows: past attempts to develop a "pure observation-language" have failed (p. 126); the attempts that come near to succeeding all presuppose some theory about what is observed (p. 127); and the hope that some future attempt will succeed rests entirely upon a questionable view of the fixity or neutrality of the content of immediate sense experience (pp. 126-7).
34. In later writings Kuhn is careful to specify that meaning variance is only partial, e.g. (1970b, p. 267). In the following

remark he claims always to have meant this: '"some difference in some meanings of some words [theories] have in common" is the most I have ever intended to claim' (in Suppe (1977, p. 506)). Yet it must be said that this was far from obvious in the original discussion in his (1970a).

35. For the charge of idealism see Scheffler (1967, p. 19); the issue is discussed at length in Nola (1980a). This and related issues will be discussed in Chapters 6 and 7.
36. Kuhn points to a parallel between incommensurability and translational indeterminacy on several occasions (e.g. 1970a, p. 202, 1970b, p. 268 and 1976, p. 191). Later, however, he distinguishes the two notions sharply (1983, pp. 679-81).
37. Admittedly, if there is a choice between incorrect translations, one might say that translation is indeterminate. But for Quine indeterminacy implies a choice between equally good translations, not a choice between equally bad ones. His point is that there are numerous translations consistent with the linguistic evidence, not that there are none.
38. For change of reference, cf. Kuhn's remark that "the line separating the referents of the terms 'mixture' and 'compound' shifted; alloys were compound before Dalton, mixtures after" (1970b, p. 269).
39. The point that more than conceptual difference is required is made with reference to Kuhn by Feyerabend (1981h, p. 154, n. 54).
40. Siegel points out that Kuhn's remarks appear self-contradictory: "unless there is a substantive difference between "comparison" and "point-by-point" comparison, Kuhn is saying that incommensurable paradigms can be compared, but not compared "point-by-point". This is equivalent to saying that they can be compared, but not compared, which does little to illuminate Kuhn's position" (1987, p. 61). Siegel is right that Kuhn's discussion is imperspicuous; yet he seemingly overlooks the "substantive difference" provided by Kuhn's explicit mention of comparison by means of reference (see next paragraph in text).
41. Kuhn's remarks about reference indicate acceptance on his part of the Scheffler-style point that reference suffices for comparison. This is further apparent in his (1979, pp. 411, 417) where, with some reservation, he endorses the causal theory of reference as a "technique for tracing continuities

between successive theories and [ ] for revealing the nature of the differences between them" (1979, pp. 416-7).

42. Kuhn notes that "the claim that two theories are incommensurable is more modest than many of its critics have supposed", and says that "insofar as incommensurability was a claim about language, about meaning change, its local form is my original version" (1983, p. 671). Suffice it to say that, while this may very well have been what he originally intended, it is not what he originally conveyed.
43. Apart from the case of phlogiston which I am about to discuss, Kuhn argues that Newtonian 'force' and 'mass' are conceptually interrelated via the second law, claiming that: "Newtonian 'force' or 'mass' are not translatable into the language of a physical theory (Aristotelian or Einsteinian, for example) in which Newton's version of the Second Law does not apply" (1983, p. 677).
44. Kuhn's holistic point seems to contrast with Feyerabend's view that terms defined using one set of principles cannot be defined by means of the incompatible principles of a rival theory. But the contrast may only be apparent: while interdefined terms cannot be introduced singly into a different holistic cluster, what presumably differentiates the clusters is incompatibility of basic principles.

2 Reference and theory comparison

2.1 Introduction

This chapter presents the approach to content comparison based on appeal to reference. The main tenet of the approach is that theories are comparable if they share reference. Its main difficulty is showing that the necessary relations of reference obtain.

The approach derives from an objection of Scheffler's to the incommensurability thesis. Scheffler criticized the close association which Kuhn and Feyerabend make between change of meaning and absence of logical relations. He pointed out that because sense may vary independently of reference, it does not follow from difference of meaning that statements are unable to contradict one another.

Scheffler's point raises the issue of how to determine whether terms from different theories refer to the same things. Scheffler himself worked with a classic description theory of reference on which a term's reference is determined by satisfaction of its associated description. However, such a theory of reference has acute difficulties in dealing with conceptual change: e.g. it implies excessive discontinuity of reference in the evolution of particular concepts and between rival concepts. Indeed, the description theory of reference even lends support to the incommensurability thesis: for theories with incompatible descriptive content would have no common reference and hence fail to enter logical conflict.

The approach was extended by Putnam, who embraced referential comparison within a causal theory of reference. Putnam suggested that such a theory of reference better suits conceptual change in science, since it allows co-reference between rival concepts and referential stability for evolving concepts. The causal theory of reference denies descriptions the reference-determining role accorded them by the description theory. It takes reference to be determined by original term-introductions in which reference is fixed ostensively or by contingent description. Accordingly, terms may continue to refer to the same thing even if their associated descriptions change; and terms may co-refer though their descriptions conflict. Hence theories may diverge conceptually even to the point of incompatible descriptive content, yet still have common reference.

While Putnam's suggestion affords real insight, the causal theory of reference faces severe problems of application in the present context. It implies too much stability: whereas the description theory leads to an excess of reference change, a causal theory which emphasizes original term-introductions effectively precludes change of reference.[1] Nor can the causal theory dispense with descriptions altogether. In the case of ostensive introduction of natural kind terms the deictic component of the act of ostension must be supplemented by at least minimal descriptive apparatus to determine the kind of thing ostended. Considerably more description is necessary to secure reference for theoretical terms whose reference cannot be picked out ostensively. Such problems of application necessitate modifications of the causal theory of reference, which reduce the impact of the causal-theoretic response on the problem of referential variance.

The topics dealt with in this chapter are arranged as follows. Section 2.2 discusses Scheffler and several issues relating to his proposal about referential comparison. In 2.3 I consider the problems which militate against the description theory of reference in connection with incommensurability. Section 2.4 introduces the causal theory of reference. Section 2.5 addresses the problem of accounting for reference change within the context of the causal theory of reference. And in 2.6 the manner in which descriptions enter into the determination of reference of theoretical terms will be discussed.

2.2 Reference and comparison

This section is concerned with Scheffler's point that statements from different theories may contradict one another as long as their terms have the same reference. Several proposed refinements of Scheffler's approach will also be considered, the common theme of which is that exact equivalence of extension is not required for comparison.

Let us recall where Kuhn and Feyerabend stand on comparison of content. As we saw in Chapter One, incommensurability is linked with irreducibility. Kuhn and Feyerabend both claim that a theory which is incommensurable with another cannot be reduced to it by deductive subsumption. For Feyerabend, incommensurable theories are "deductively disjoint" (1978, p. 67) in the sense that they do not share common consequences. Their terms differ in meaning, so no statement derived from one theory is either asserted or denied by the other; their "content classes... are incomparable" (1975, p. 223). For Kuhn, similarly, statements from incommensurable theories cannot be expressed using the same vocabulary: "There is no neutral language into which both of the theories as well as the relevant data may be translated for purposes of comparison" (1979, p. 416). Admittedly, Kuhn has come to accept comparison by means of reference (see 1.3); but, given his denial of translation, he remains committed to inability to compare theories by means of statements expressed in a shared and semantically invariant vocabulary.

Incommensurable theories do not permit comparison with regard to specific points of disagreement, as they are unable to assign conflicting truth-values to a set of shared statements with common meaning. Feyerabend is explicit about this in his concession of Shapere's point that theories without common meaning cannot contradict one another; and Kuhn has the same thing in mind when he says incommensurable theories cannot be translated into a common language in which they may be compared in "point-by-point" manner.[2] It is precisely this link between difference of meaning and inability to contradict which is put in question by Scheffler.

In his (1967, pp. 58-61), Scheffler criticized the inference from change of meaning to incommensurability, arguing that such change does not preclude the semantic relationships necessary for theory comparison. He makes essentially two points: that terms may differ in sense and still have the same reference; and that such sameness of reference is a sufficient semantic condition for statements to enter logical relations such as contradictoriness.

Thus, while Scheffler grants that sense may vary with theory, he denies that such change of meaning is invariably accompanied by change of reference:

> Terms may denote the very same things though their synonymy relations are catalogued differently. Hence common reference may ... survive synonymy alterations bearing directly on the very terms in question. Opposing theorists may differ in respect of these [ ] alterations ... they may yet mean, that is, refer to, the same things. (1967, p. 60)

Given the possibility of stable reference, variation of sense does not itself entail the absence of logical relations. For, as Scheffler notes,

> deduction within scientific systems ... requires stability of meaning only in the sense of stability of reference. That is to say, alterations of meaning in a valid deduction that leave the referential values of constants intact are irrelevant to its truth-preserving character. (1967, p. 58)

Since "disagreement in the sense of explicit contradiction" requires only sameness of reference, stability of reference "reinstates the possibility of such disagreement, itself involved in any plausible conception of rational discussion" (1967, pp. 60-1).

What underpins Scheffler's contention about the possibility of contradiction is this: the truth of a sentence may preclude another sentence's being true even if their constituent terms differ in sense. In extensional contexts, reference rather than sense is the semantic property crucial to determining truth. The truth-value of a sentence depends on the referential properties of its component expressions; e.g. a simple predicative sentence is true if and only if the object denoted by its subject-expression is in the extension of its predicate. Analogously, a pair of sentences may be referentially so related that both cannot be true; e.g. if one is a negation and the other a positive assertion, and if the reference of the component expressions of one is identical with that of the correlative expressions of the other, then if one is true the other is false.[3] Since it is referential properties which are of relevance to the truth of sentences, differences of meaning not reflected by difference of reference do not affect the properties of sentences associated with truth. Thus, Scheffler's argument reveals how theories whose terms differ in meaning may stand in conflict on particular points.

Before turning to extensions of Scheffler's approach, I will consider a pair of objections that might be brought against it. The first objection stems from the point that contradiction requires more than identity of reference of sentence components. Scheffler's core idea is that as long as they contain co-referential terms, contradictory sentences need neither be formed out of expressions the same in meaning, nor of words with the same physical form.[4] As against this, however, it may be argued that contradiction in a strict sense is not merely an extensional relation. For formal contradiction, a negation must negate a sentence which is a token of the same sentence-type as the sentence it contradicts.[5] Thus, the merely extensional relationships appealed to by Scheffler do not suffice for the possibility of contradiction in a strict formal sense.

However, this objection does not present any serious threat to Scheffler's approach. Sentences need not contradict one another in a strict formal sense in order to be incapable of jointly being true. Thus, it suffices to respond that, even if sentences with merely co-referential components cannot formally contradict one another, it may still be the case that both of the sentences cannot be true. Identity of reference of sentence components is enough for the truth of one to be able to preclude the truth of the other. And such conflict between the sentences of rival theories is enough for comparison with respect to particular points of disagreement.

The second objection stems from the absence of logical relations across languages. Without being translated into a common language, sentences from different languages cannot enter logical relations (e.g. contradiction, entailment) with one another. For the result of linking such sentences together would not be a syntactically well-formed formula of either language. Since incommensurable theories purport to be expressed in different and untranslatable languages, no sentence of one may bear a logical relationship to a sentence of another. So even if their terms do share reference, Scheffler's argument is beside the point.

As before, there is no need to question the premise of this objection. It may be granted that logical relations do not obtain directly between the sentences of different languages. But from this it does not follow that sentences of different languages are unable to be related in such a way that both cannot be true.

More to the point, the objection exaggerates the sense in which incommensurable theories are expressed in different languages. Kuhn and Feyerabend do not identify the languages of incommensurable theories with distinct natural languages; rather, their concern is with semantic relations between the vocabularies used by particular theories. The vocabulary employed by a theory amounts, at most, to a part of a natural language. As such, it constitutes a sub-language or a local idiom within a larger, encompassing language. Since such sub-languages are embedded within a background natural language, no linguistic barrier prevents logical relations between them. What does, however, preclude logical relations between theories, according to Kuhn and Feyerabend, are the semantical differences between the vocabularies employed by such theories. In sum, incommensurability is a semantical relation between the vocabulary specifically employed by theories; it is not a relation between total natural languages.[6]

Neither objection poses a threat to the referential approach to theory comparison, so let us proceed with extensions of Scheffler's position. We may take Scheffler to have shown that the content of theories with common reference is comparable. For sentences from rival theories may be compared with regard to conflict provided only that their constituent expressions have identical reference. The refinements I am about to consider suggest that even weaker relations than identity of reference suffice for such comparison.

Scheffler's own treatment of comparison involves sameness of reference. But it seems plausible that comparison may proceed by means of referential relations other than identity. Scheffler's approach was first extended along these lines by Martin (1971 and 1972), who argued that relations of extensional overlap also make conflict possible. Reference need not be identical, Martin notes, for if the extension of one predicate contains the extension of another as a subset, an assertion made using the one may be inconsistent with a denial which uses the other (1971, p. 25).[7] Martin also shows that conflict can occur if extensions intersect in the sense of sharing a subset: if the extensions of predicates intersect without containment, statements made with the predicates may be inconsistent with respect to the objects in the intersection (1971, pp. 25-6). Thus, in general, if terms from rival theories overlap in the sense either of extensional containment or intersection, then it is possible for their consequences to conflict.

A similar point about comparison by means of extensional overlap may be derived from Field's notion of "partial denotation". In his (1973), Field argues that terms may undergo "denotational refinement" in the sense that their extensions may be narrowed down or divided up in the transition to later theory. The way in which an earlier term refers may fail to discriminate between entities which are subsequently distinguished, so that its reference is indeterminate between them. In such cases, Field says that the term "partially denotes" both entities. Thus, he claims, for example, that Newton's term 'mass' partially denoted relativistic and proper mass (1973, p. 476), and that it underwent denotational refinement in the transition to relativity theory (1973, p. 479).

The notion of partial denotation suggests various ways for terms to overlap extensionally without having exactly the same reference. Accordingly, sentences whose terms partially refer to the same things are able to conflict with one another. For suppose that a term's extension is refined in such a way that it comes to refer to a proper subset of its original extension; i.e. it formerly referred partially to that subset as well as to some other. In such a situation, statements made prior to the refinement may still conflict with respect to the properties of the common subset with statements made after the refinement. Thus conflict and comparison of sentences may be based on shared partial reference in the absence of sameness of reference. It should be noted, however, that the comparability afforded by partial co-reference is formally the same as that provided by the extensional overlap discussed by Martin. For if two terms partially refer to the same thing, then their extensions intersect. So, as far as content comparison is concerned, Field's partial co-reference and Martin's extensional overlap come to the same thing.

A distinct form of reference overlap has been discussed by Kitcher (1978).[8] He argues that some tokens of a given term-type may refer to different things than do other tokens of the same term-type.[9] One of Kitcher's main examples is the phlogistic expression 'dephlogisticated air'. He claims that certain tokens of the expression were used to refer ostensively to samples of oxygen gas, while other tokens failed to refer. Such differential reference of tokens of a term-type suggests the possibility of inter-theoretic co-reference at the level of tokens. Tokens of different term-types employed by rival theories may refer to the same things, permitting theory comparison to be made on the basis of such tokens. For example, if certain tokens of 'dephlogisticated air' co-refer with certain tokens of 'oxygen', then statements made using the tokens of the former expression may agree or disagree with statements which use tokens of the latter.

In sum, Scheffler's appeal to sameness of reference and the modifications proposed by Martin, Field, and Kitcher show how theories may be compared in various ways by means of reference. So even if meaning does vary with theory, as Kuhn and Feyerabend maintain, the extensions of terms from rival theories may still contain common members, with respect to which comparison can be made.[10]

2.3 Conceptual change and the description theory of reference

The considerations so far advanced in favour of the referential approach yield at most a conditional result: if the terms of rival theories overlap extensionally, then there is a basis for comparison of content. Thus, the question remains whether the necessary relations of reference actually obtain. Consideration of this issue leads into the theory of reference.

The problem initially facing the referential approach is that the possibility of non-synonymous co-referential expressions does not itself warrant the presumption of referential stability through alteration of sense. Alteration of a term's sense can just as well be accompanied by change of reference.[11] Thus, to note simply that terms which do not mean the same may co-refer does not rule out reference changing along with sense. So while the possibility of non-synonymous co-referential expressions precludes the inference from meaning variance to incomparability, it does not itself constitute evidence in favour of referential stability.

Nor does Scheffler's own treatment of reference provide any guarantee of cross-theoretic reference. He does claim that reference change is less plausible than change of sense:

> The insistence ... that theoretical incorporation affects meanings is plausible at best only with respect to senses, and even so only for certain theoretical incorporations ... Such alteration ... does not automatically effect a disruption of referential constancy. (1967, p. 62)

But while this may be true, if only because reference change is not entailed by change of sense, it follows neither that reference is unable to change with change of theory, nor that it tends not to change.

Scheffler also suggests that empirical "laws may retain their referential identities throughout variations of theoretical context" (1967, p. 62). He supports the point by noting that the reference of terms appearing in such laws can "be determined independently of a characterization of their respective senses" (p. 61), so that their "constancy of referential interpretation is ... accessible to reinforcement through shared processes of agreement in particular cases" (p. 62). Yet this only shows that constancy of reference can be established without considering sense, which in itself shows nothing about the extent to which such constancy prevails in the transition between rival theories.

The issue is of fundamental importance, since neither Kuhn nor Feyerabend restrict the semantic differences attendant upon incommensurability to variation of sense. Both authors take the transition between incommensurable theories to involve changes of reference. Indeed, for Feyerabend and the early Kuhn, such transitions result in a radical discontinuity of reference. On Feyerabend's account, conflict between the defining principles of the concepts of incommensurable theories is such that their terms cannot have common reference.[12] Similarly, on Kuhn's original account conceptual difference between paradigms leads to wholesale change of reference from the "world" of one paradigm to the "world" of another.[13] In contrast, Kuhn's more moderate later view involves localized classificational changes which take place against a background of overall referential stability. While theories may purport to refer to different entities, Kuhn lays primary emphasis on extensional changes brought about by the redistribution of sets of objects among "taxonomic categories".[14]

The thesis that reference changes in the transition between conceptually disparate theories raises the issue of how reference is determined; i.e. of what it is that determines that a term has the reference it does have. According to Kuhn and Feyerabend, reference varies because the concepts employed by different theories vary. This suggests a view of reference on which what a term refers to depends on the concept it expresses, and change in the concept expressed can result in change of a term's reference. Thus, notwithstanding the possibility of non-synonymous co-reference, sufficient difference in the concepts employed by theories could prevent their terms from having common reference.

The dependence of reference upon the concept expressed by a term is a characteristic tenet of the description theory of reference, according to which the reference of a term is determined by its associated descriptive content. On such an account of reference, the conceptual content or sense of a term is expressed by a description which specifies properties that identify the term's reference. An object or set of objects which satisfies the description by virtue of possessing the specified properties qualifies as the referent. Such satisfaction of descriptive content is both necessary and sufficient for reference: an object which satisfies the description is the referent and one which fails to satisfy it is not. More generally, satisfaction of a term's associated description is necessary and sufficient for having reference: if something satisfies its description, then the term has reference; if nothing satisfies it, the term fails to refer. Because satisfaction of the description is a necessary condition for reference, a term introduced to refer to a particular object may fail to refer to it if the term's associated description does not specify the intended referent correctly.[15]

On this account, a term's reference can change if its descriptive content is altered in such a way that it is no longer satisfied by the same thing. This accords with Kuhn's and Feyerabend's emphasis on conceptual disparity in their analysis of reference change.[16] But it is not only their analysis which presupposes a description theory of reference. Scheffler too, in spite of being opposed to their views, is committed to such an account of reference. His point about co-reference without synonymy explicitly utilizes a Fregean dichotomy of sense and reference, according to which reference is determined by sense in the manner specified by the description theory (see Scheffler 1967, pp. 54-6). There is, however, a basic problem with such an appeal to the description theory, which is evident in the first instance from Kuhn's and Feyerabend's own reliance on it. Namely, the conceptual differences at issue may be unable to be construed as correct alternative descriptions of the same things.

The problem stems from what is in effect a corollary of the description theory with respect to co-reference. The theory readily accounts for the phenomenon of non-synonymous co-reference: the same objects may be described in non-equivalent ways, so the same reference may be determined in different ways. But there are limits to the determination of common reference by unlike descriptions. It follows from the requirement that reference must satisfy description that descriptions must be jointly satisfied by the same objects in order to pick out the same reference. If descriptions are so related that they cannot be true of the same things, then the terms for which they determine reference cannot co-refer. Consider, for example, the case of a term retained throughout the evolution of its associated concept. If properties formerly taken to determine the term's reference come, as a result of that evolution, to be denied of its referent, then the term would be unable to preserve its reference. Similarly, terms from rival theories which are associated with incompatible descriptive content would not have their reference determined by descriptive specification of the same thing.

Kuhn and Feyerabend discuss numerous cases of conceptual disparity which do not lend themselves to analysis as alternative non-equivalent descriptions of the same thing. An example of a term retained through conceptual change discussed by both authors is the term 'mass'. They both point out conflicts between the descriptive content associated with the term as it occurs in Newtonian and Einsteinian physics. Kuhn notes that "Newtonian mass is conserved; Einsteinian is convertible with energy" (1970a, p. 103), while Feyerabend remarks:

> That the relativistic concept and the classical concept of mass are very different indeed becomes clear if we also consider that the former is a relation, involving relative velocities, between an object and a coordinate system, whereas the latter is a property of the object itself and independent of its behaviour in coordinate systems. (1965, p. 169)

As for different terms used in rival theories, Feyerabend asserts that 'impetus' is not co-extensive with 'momentum': "whereas the impetus is supposed to be something that pushes the body along, the momentum is the result rather than the cause of its motion" (1981d, p. 65). Kuhn discusses a more extreme case involving key concepts of phlogistic chemistry (1983). Owing to differences of ontology, phlogistic concepts cannot be understood as mere alternative descriptions of the same entities as those described by the oxygen theory. Since the oxygen theory rejects phlogiston, the descriptive content of terms definitionally linked with 'phlogiston' is incompatible with the oxygen theory.

Examples such as these are not readily construed as jointly satisfiable descriptions. Thus, they expose the basic problem facing the attempt to found the referential approach to content comparison on the description theory of reference. Namely, an important kind of conceptual change dealt with by Kuhn and Feyerabend resists analysis in terms of the model of alternative descriptions of the same thing, and so appears to involve failure of co-reference.[17]

The problem is, however, more general than this would suggest. The description theory of reference leads to a thesis of referential discontinuity in its own right. On the assumption that reference is determined by description, significant conceptual changes or differences may be presumed to yield difference of reference. For in the transition between theories, if a theory's descriptions of its purported objects of reference were replaced by incompatible descriptions or by descriptions of completely different objects, at least some of the terms used by the theories would fail to have common reference.

This theme is found in a number of authors,[18] most notably Putnam, who writes that:

> Bohr assumed in 1911 that there are (at every time) numbers p and q such that the (one dimensional) position of a particle is q and the (one dimensional) momentum is p; if this was part of the meaning of 'particle' for Bohr, and in addition, 'part of the meaning' means 'necessary condition for membership in the extension of the term', then electrons are not particles in Bohr's sense, and, indeed, there are no particles 'in Bohr's sense'. (And no 'electrons' in Bohr's sense of 'electron', etc.) None of the terms in Bohr's 1911 theory referred! It follows on this account that we cannot say that present electron theory is a better theory of the same particles that Bohr was referring to. (1975a, p. 197)

Here Putnam shows how a strict reliance on descriptions would prevent reference from being secured by mistaken characterizations of intended objects of reference. The result in the case of the evolution of concepts, as with that of the electron, is an excessive instability of reference. Instead of such strict reliance on descriptions, therefore, what is called for in the context of scientific conceptual change is a theory of reference which is less sensitive to variation of descriptive content.

2.4 The causal theory of reference

To apply the referential approach in the context of significant conceptual variance, it needs to be shown that reference is able to survive substantial change of descriptive content. Thus Putnam, having noted in the passage quoted above that the description theory leads to undue variability of reference, proceeds to advocate a causal account of reference. The leading idea of such an account is that reference is established at original term-introductions which determine reference for subsequent use; so that reference originally so determined is not affected by variation of descriptive content. This account of reference will be outlined in more detail shortly, but first I will discuss problems with the descriptive determination of reference which motivate the causal theory of reference.

Quite apart from leading to undue referential variance, the idea that satisfaction of description is necessary and sufficient for reference seems itself mistaken. In the first place, it appears not to be a necessary condition because reference may succeed even if the referent is misdescribed. The basic point here is that describing objects is an intrinsically fallible enterprise. It is possible to be mistaken about or to be ignorant of the properties of an object which one describes. This may lead one to misdescribe the object, yet it may still be clear which object is misdescribed. Even though the object fails to be correctly described it is still that object to which reference is made.

Kripke and Putnam discuss numerous examples in which reference succeeds despite erroneous description. It could turn out, they maintain, that a natural kind fails to have the properties normally attributed to it. Natural kinds may have abnormal members which fail to have properties normally associated with the kind (e.g. tigers may be albino or three-legged). More importantly, a natural kind may itself fail to have the properties it is thought characteristically to possess. Thus Kripke claims that gold could turn out not really to be yellow:

> Suppose there were an optical illusion which made the substance appear to be yellow; but, in fact, once the peculiar properties of the atmosphere were removed, we would see that it is actually blue. (1972, pp. 315-6)

In such a situation, Kripke suggests, 'gold' would still refer to the stuff wrongly taken to be yellow: "though it appeared that gold was yellow, in fact gold has turned out not to be yellow, but blue"

(1972, p. 316). Similarly, Putnam considers the case in which cats are not animals but robots, noting that in such a situation "we should continue to call these robots that we have mistaken for animals and that we have employed as house pets 'cats,' but not 'animals'" (1962, p. 660). In examples such as these reference succeeds even though one or more property is mistakenly taken to be characteristic of the kind referred to.

A similar point lies behind Donnellan's well-known distinction between referential and attributive uses of definite descriptions (1977). According to Donnellan, what is referred to by means of a definite description is determined differently, depending on the manner in which it is used. In referential use, a definite description is employed for the purpose of referring to some particular object, whether or not the object fits the description. In attributive use, a definite description is used to refer to whatever it is that satisfies it.

Donnellan gives as example the sentence "Smith's murderer is insane" (1977, pp. 46-7). The description 'Smith's murderer' occurs attributively if it is used to pick out whoever murdered Smith, without having any particular person in mind. It is used referentially if it is meant to pick out some particular person, even if it should turn out that that person is not the murderer. Descriptions which occur in referential use further illustrate the point that reference may succeed even if the object of reference is misdescribed, so that satisfying a description is not necessary for reference.

Not only is satisfaction of description unnecessary for reference, it fails also to be sufficient. Just as one may misdescribe an object to which reference is made, the converse is true. A description may be satisfied by something which is not the reference. Properties which are erroneously believed to be characteristic of a referent may in fact be instantiated by something entirely different which does not, simply by virtue of instantiating the properties, constitute the referent.

Thus Kripke, continuing his discussion of the situation in which gold is blue, notes that a substance which did satisfy the ordinary description of gold would not in such circumstances be gold:

> ... we use 'gold' as a term for a certain *kind* of thing. Others have discovered this kind of thing and we have heard of it. We thus as part of a community of speakers have a certain connection between ourselves and a certain kind of thing. The kind of thing is *thought* to have certain identifying

marks. Some of these marks may not really be true of gold. We might discover that we are wrong about them. Further, there might be a substance which has all the identifying marks we commonly attributed to identify the substance of gold in the first place, but which is not the same kind of thing, which is not the same substance. We would say of such a thing that though it has all the appearances we initially used to identify gold, it is not gold. (1972, p. 316)

Gold, therefore, is not simply to be identified as that which fulfils some standardized description. To be gold, for Kripke, it is not enough for a substance simply to satisfy such a description; in addition, it must stand in the right sort of relation to the use of the word 'gold'.

Suppose, to the contrary, that satisfaction of description were sufficient for reference. This would entail reference to whatever objects happen to satisfy a description. Thus, even if reference to something quite different were the manifest intention, what actually satisfies the description would necessarily be the referent. But such strict adherence to the letter of a description would have the effect of imposing reference to unintended objects upon speakers through the mere accident of misdescribing the preferred referent. It seems, rather, that in the same way that one may succeed in referring to a misdescribed object, one need not necessarily refer to what the description actually fits.

The point receives further support from Putnam's science-fiction example of Twin-Earth (1975b, pp. 223f). Twin-Earth, as the story goes, is a planet like the Earth in every respect but one. The difference is that what fills the lakes and seas of Twin-Earth is not H_2O but XYZ. The substance XYZ is chemically distinct from H_2O but shares its surface features. It too is clear, thirst-quenching and odourless. Putnam stipulates that the inhabitants of Earth have Twin-Earth counterparts psychologically indistinguishable from themselves. In particular, Twin-Earthians and Earthians associate the same properties with the substance they call 'water'; inhabitants of both worlds describe water as a drinkable, colourless liquid. Yet when Twin-Earthians use the term 'water' they do not refer to the same stuff as we on Earth do. For the extension of 'water' as used on Twin-Earth is XYZ, while the extension of 'water' on Earth is H_2O.

Putnam primarily intends the example to show that psychological state does not determine reference. Speakers may be in the same internal mental state yet fail to co-refer because they are extrinsically related to different things. However, the example also shows that mere satisfaction of description is insufficient for reference.[19] The substances called 'water' on Earth and Twin-Earth have the same surface characteristics, hence satisfy the same description of water. Yet, as the term is used in each of the two worlds, it refers to distinct substances. Even though XYZ satisfies the description associated with the term 'water' as used on Earth, that does not itself suffice for the extension of our term 'water' to include XYZ. Merely satisfying the description is not enough, since that would entail reference to a substance completely isolated from the linguistic practices in question.[20]

If satisfaction of a term's associated descriptive content is neither necessary nor sufficient for reference, then reference must be able to be secured in some more direct way. This is a consequence of both of the two main points which have just been discussed. If reference may succeed despite misdescription of the referent and the referent need not be what satisfies the description, then it follows that reference is at least partially independent of description. In short, this suggests that reference may be secured by means of the pragmatic relations of a non-linguistic, causal nature into which speakers enter with their environment. This more direct manner of reference determination is what the causal theory of reference attempts to supply.

Such independence of reference from description implies greater stability of reference through variation of descriptive content. Thus Putnam, in advocating a causal view of reference in his (1975a), supports both the independence of reference from description and the continuity of reference through conceptual change. The shift from the description to the causal theory of reference involves a fundamental shift of perspective. It requires us to turn our attention from the descriptions which speakers currently associate with a term to the circumstances under which the term was originally introduced as a name for some object or kind.

In outline, the manner in which reference is determined according to the causal theory is as follows. Initially, a speaker introduces a term into the language by naming an object or kind of object at an informal naming ceremony. At the introduction of the term the object or kind named is singled out by ostension or by a description. In subsequent use the term continues to refer to the entity to which it was originally attached on the occasion of its introduction. Speakers who acquire the use of a term at a naming ceremony pass it on to other speakers, who in turn pass it on to others. In this way, later use of a term inherits reference from earlier use. Speakers not present at a naming ceremony acquire the term from other speakers, inheriting the reference via a chain of communication which extends back to the original introduction of the term. On this picture of reference, to find out what a term refers to, the use of the term is traced back to its initial use and its reference is what was singled out for naming in the original ceremony.

The key element in this account is the manner in which the referent is originally singled out. Here the simple case is the naming of a particular individual (e.g. a person) by direct ostension. In such a case, the object to be named is indicated by means of a demonstrative device (e.g. pointing) and the name is given to the object indicated. What fixes the object as the referent is the causal relation of perceiving the object, irrespective of any description offered by speaker or audience.

The reference of certain general terms is also fixed ostensively on this account. For example, the reference of some natural kind terms is established by ostension of a sample of the natural kind. Since the entire kind cannot be present at the ostension, the extension of the term is fixed by means of a representative sample. The extension is the kind instantiated by the sample: it consists of the set of objects which bear the same-kind-as relation to objects in the original sample. Typically, this is a theoretical relation, in the sense that it is determined by internal structural traits which require scientific research to discover. Even if an object has the same surface traits as the sample members of the kind, it cannot be a member of the same kind unless it has the same internal structure as the sample objects. This explains why the XYZ in Putnam's Twin-Earth story is not water: since the samples of water used to fix the reference of our term 'water' have molecular structure H_2O, the extension of the term is H_2O and does not include XYZ.[21]

Not all terms have referents which may be picked out ostensively, however. Terms which refer to unobservable objects do not admit of ostensive introduction. With such terms, descriptions play a role in securing reference, though not the same role as that accorded them by the description theory. There may be circumstances, suitable causal relations not obtaining, in which an introductory description occurs in attributive use in Donnellan's sense, so that it specifies only such entities as satisfy the description, and nothing else. But this is not the only possibility: a description may be used to specify non-ostensive reference at a naming ceremony without being taken as a statement of necessary or sufficient conditions of reference.

Because of the relative likelihood of misdescribing an unobservable referent, causal considerations tend to take precedence in determining what is picked out by the use of a description at a naming ceremony. This takes into account the possibility of securing reference by the use of a description even if the referent is misdescribed. For example, if the description occurs referentially in Donnellan's sense, then the reference may be determined by the speaker's causal relationship with the intended referent. Alternatively, an unobservable referent may be specified by its observable effects. As Kripke and Putnam suggest, entities such as heat and electricity may be picked out by causal descriptions identifying them as the entities causally responsible for specific observable effects.[22] Here too reference is fixed by causal relationship: it is the causal relationship between the observed effects and the entities responsible for the effects which fixes the reference.

The causal ingredient in this account is due to its emphasis on the relations speakers have to other speakers and their environment rather than on the way referents are described. It reflects the view that the causal and pragmatic engagement of speakers with the items about which they speak, as well as with each other, are crucial in determining what they refer to.[23] As such, causal relations enter the account at two key points: the relations of speaker and reality, and the relations between speakers. The first involves causal relations between speaker and referent in a naming ceremony. The second involves the acquisition of reference by speakers absent from a naming ceremony by a chain of communication linking their use with original use. In neither the original nor in subsequent use need a speaker associate a description with the term which correctly or uniquely identifies the referent, for if suitable causal relations obtain the speaker may use the term to refer.

Such an account frees speakers of the need to associate the same descriptive content with a term throughout its use. To preserve reference, later speakers need not associate the same descriptions with a term as those present at its introduction. Indeed, if they acquire the use of a term from others who are linked to the introduction of the term by a communicative chain, later speakers need not associate any particular description with a term. Provided they are suitably linked to initial use, speakers may continue to refer to the referent established at the introduction of a term, even if the descriptive content they associate with the term is incompatible with that which speakers involved in its original introduction associated with it.

The possibility of referring to the same thing despite disparity of descriptive content is the basis of the causal theory's response to the thesis of referential incommensurability. Since, according to the causal theory, the reference of terms which occur in scientific theories is normally established at their original introduction, the scope for continuity of reference through conceptual change is considerable. For if the reference originally established for a term is preserved despite variation in the properties taken as characteristic of the referent, descriptive specification of the referent may change without altering reference.

Consider first the case of particular concepts which undergo change. Two examples of such concepts mentioned in the preceding section were Newton's and Einstein's rival notions of mass, and Putnam's case of the Bohr electron. In the case of the evolution of such concepts, the causal theory avoids the excessive referential instability of the description theory. For the reference of the term which expresses the concept is not liable to change with significant change of concept: even if properties initially attributed to the referent are later denied as a result of changes in the concept, the reference may be preserved. Since the concept does not determine the reference, the term continues to refer to the referent established at its original introduction.

Similarly, widespread conceptual change in the transition between theories need not result in radical change of reference. As was noted at the end of the preceding section, the description theory of reference invites the presumption of referential discontinuity in the case of extensive conceptual change. Rival theories which specify their objects of putative reference by means of incompatible descriptions, or which describe disparate sets of objects, would be unable to share common reference. However, profound conceptual change affecting the language retained in the transition between successive theories does not have this outcome according to the causal theory. Since reference is established independently of the way in which referents are described, major transformation of conceptual apparatus does not entail radical discontinuity of reference between the theories.

Such consequences are not limited to continuity of reference for terminology retained in the transition between theories. Similar remarks apply to the case in which distinct terms expressing unlike concepts are employed by rival theories in the same contexts. Examples of such concepts are those of oxygen and dephlogisticated air, and impetus and momentum. The reference of such terms is determined by what they were applied to on the occasion of their introduction and not their associated descriptive content. Thus, even if the concepts they express are in conflict, a common referent may still have been secured in their original introduction.

2.5 Initial term-introductions and reference change

The causal theory of reference supports the view that concepts from rival theories may differ and individual concepts undergo change, yet still be concepts of the same things.[24] It also supports Scheffler's point that theories may be compared via reference overlap by extending his position to cover the case of terms associated with jointly unsatisfiable descriptions. Since it makes reference depend on the way a term is first introduced, the causal theory divorces reference from the changing descriptive associations which speakers connect with the terms they use. Thus, it provides an account on which reference is less sensitive to variation of descriptive content, the need for which was noted at the close of section 2.3.

The danger is that the causal theory may be too insensitive. It seems to rule out the possibility of reference change altogether. As has been argued by Fine (1975), if a term's reference is established on the occasion of its introduction, it cannot change. The problem with the causal theory, according to Fine, lies with its reliance on initial term-introductions:

> In the beginning of the use of a term (like 'water', 'compound', 'electron', etc.), the term is attached by an act of introduction (conventional definition, ostension, or whatever) to some existent. Thereafter, the term refers to that existent to which it was originally attached. (1975, p. 23)

For example, to find out what 'compound' refers to, "trace back the chain of uses of 'compound' to the introductory act, and then 'compound' refers to whatever was picked out on that occasion" (p. 23):

> Because 'compound' attaches directly to its referent, there is no possibility that the referent could change its station as referent over time. (1975, p. 24)

In short, the idea that a term's reference is fixed once and for all at its first use renders reference change impossible. To the extent, therefore, that the causal theory is committed to such a reference-establishing role for initial term-introductions, it implausibly precludes change of reference.

Apart from undue preclusion of reference change, the idea of an initial term-introduction at which reference is established is itself objectionable. No doubt terms generally do have a first occasion of use. And no doubt reference may be established on such occasions. But the picture of present referential use as ultimately deriving from an original term-introduction is idealized. To assume that a term's reference is established only once and thus that one introduction underlies all subsequent referential use of a term is just to assume that matters are less complicated than they might be.

The idea of an original term-introduction is a simplification, useful in explaining the role of naming ceremonies as against associated descriptions. But its applicability to actual linguistic use cannot be taken for granted. The main problem is that a term might originally be introduced to refer to one kind, yet subsequently, whether by accident or by deliberate choice, it may come to be applied to another kind. Later use might then be based on subsequent application. For example, the use instituted at the original introduction might be discontinued and the term be re-introduced as a name for the kind subsequently picked out. It might even happen that later use of a term traces back to more than one term-introduction, with the possible result that the term is linked to different kinds of things. Thus, so far from showing that determinate reference to particular objects or kinds is guaranteed by the role initial term-introductions play in fixing reference, the assumption that reference is fixed at an original naming ceremony simply presupposes it.

The notion that reference is fixed at an initial term-introduction not only prevents change of reference, it is based on an unrealistic picture of the role of initial use in establishing reference. However, neither of these problems strikes at the heart of the causal theory of reference, for the idea that reference is inalterably fixed by first use is an inessential part of the theory. To allow for reference change and remove the assumption that reference is fixed from the start, it suffices to grant subsequent use a role in determining reference. If applications of a term on other than the occasion of its inaugural use have the ability to affect its reference, then applying the term under circumstances which differ from those obtaining at the time of its introduction can result in alteration of reference.

Allowing subsequent use a role in determining reference represents no departure from the causal theory of reference. For even if reference need not be fixed solely by initial term-introductions, it may still be fixed by means of causal relations linking speakers with items in their environment. And reference may still be transmitted from earlier to later use by a chain of communication without a need for later speakers to associate the same descriptions with terms as earlier speakers. The role of subsequent use in determining reference has been incorporated into the framework of the causal theory by Devitt and Kitcher, whose ideas I will now discuss.

Devitt responds to Fine's objection to the causal theory by pointing out that terms may be "multiply grounded" in their referents.[25] For Devitt, a term is "grounded" in its reference by means of a causal (normally perceptual) link with it.[26] Such a causal link is what fixes the term's reference in either an original naming ceremony or in subsequent application of the term to its referents. Thus, in saying that terms are multiply grounded, Devitt means that there may be multiple causal links directly attaching them to their referents:

> ... the causal networks underlying natural kind terms are usually multiply grounded in objects... The act of introduction is only one of many confrontations between a term and the world. Given the fact of multiple groundings we can hope to explain reference change by finding changes in the pattern of groundings over time. (1979, p. 41)

If a term is grounded by multiple causal links with its referents, the possibility arises that it is grounded in different kinds of things. For example, it could occur that a term is grounded uniformly in samples of one natural kind prior to a given time, while all groundings of the term after that time are in samples of a distinct kind. Alternatively, a term might be grounded initially in samples belonging to two or more kinds, and later be grounded in only one of the kinds represented by the original sample.

Such shifts in the "pattern of groundings" of a term constitute a change of reference.

Devitt's idea that later applications of a term can also ground it resolves the difficulties associated with fixing reference exclusively at initial term-introductions. For not only does it meet Fine's objection, it dispenses with the unlikely assumption that all subsequent referring uses of a term derive their reference from a single naming ceremony. However, Devitt conceives grounding primarily on the pattern of ostensive term-introduction, in which a term is attached to its referent by means of direct perception.[27] The trouble is that the model of ostensive term attachment is not readily applicable to theoretical terms. As we will see in the next section, the causal theory must grant an extensive role to descriptions in the determination of reference for theoretical terms. Thus, given that Devitt's notion of grounding is so closely modelled upon direct ostension, his idea of multiple groundings cannot be applied in unmodified form to theoretical terms.[28] For this reason I now turn to Kitcher, whose approach to the reference of theoretical terms does have a role for descriptions.

Kitcher's view that different tokens of a term-type may refer differently was dealt with briefly in section 2.2, where it was considered as a variant of Scheffler's point that co-reference suffices for comparison. The view forms part of Kitcher's thesis that different tokens of the same scientific term-type may have their reference fixed in different ways. Briefly, his thesis is that a term may attach to its referent in more than one way, since it may be applied in different contexts as well as to novel instances; use of the term appears unified to its users because they believe the different ways in which its reference is fixed to be alternative ways of securing reference to the same thing.

Kitcher supports this thesis with analysis of cases of conceptual disparity of the sort which first suggested incommensurability. A major source of his examples is the transition from phlogistic to oxygen chemistry.[29] He argues, for instance, that when Priestley applied the term 'dephlogisticated air' to samples of oxygen gas, its reference ceased to be fixed solely by a description based on the definition of 'phlogiston' and became attached as well to oxygen:

> Many tokens of this term produced by Priestley and other phlogistonians have their referents fixed through a causal chain initiated by the event in which Stahl explicitly

> specified phlogiston as the substance emitted in combustion. These tokens fail to refer. But, after Priestley had isolated oxygen and misidentified it, many subsequent tokens of 'dephlogisticated air' had their reference fixed through causal chains initiated by encounters with oxygen. Those tokens refer to oxygen. Phlogistonians engaged in linguistic practices which enabled them to produce tokens of 'dephlogisticated air' initiated in these two different ways because they were confident that the gas isolated from the red substance obtained by burning mercury was dephlogisticated air. (1983, p. 696)

By an event which initiates a causal chain, Kitcher means an event in which a term's reference is fixed by description or ostension, and which stands at the beginning of a chain of communication linking earlier with later use. So, on his analysis, the reference of 'dephlogisticated air' is fixed in two ways, by a phlogistic description and by ostension of oxygen gas. From the perspective of phlogistic chemists both ways secure reference to the same substance, so they take tokens of the term deriving from either way of fixing reference to be co-referential. From a later perspective, however, the two ways of fixing reference fail to pick out the same thing, so tokens of the term which derive from the two different ways of fixing reference appear not to co-refer.

Kitcher's analysis yields a picture of the use of scientific terms on which their reference may be established on more than one occasion, often in unlike ways. Yet, this diversity will seem unexceptionable to the users of such a term given their belief in the unity of reference underlying its use:

> From the perspective of the users of the terms, the multiplicity of initiating events is likely to seem unproblematic (even if it should be explicitly recognized). Some of the initiating events are observations and experiments involving a particular kind or thing, and the event serves to fix reference to that kind or to that thing. Others are events in which an object, or set of objects, is singled out by description. For the users of the term, these events are taken to identify the same referent. They believe that the same kind or thing is present on all the occasions of observation and experiment, that the same kind or thing is singled out by all the descriptions. (1983, pp. 695-6)

Since terms used in science are constantly being applied to novel instances and in various contexts, such differential reference-fixing of tokens of a term is according to Kitcher commonplace and even unavoidable.

In connection with incommensurability, the primary aim of Kitcher's approach is to show how the languages of conceptually disparate theories may be linked referentially.[30] For, as I noted in 2.2, terms from rival theories may be unable to co-refer as term-types, yet some of their tokens may refer to the same things. This approach also provides an account of reference change. For as the set of ways in which a term's reference is fixed undergoes change, so too may the term's reference vary. As with Devitt, Kitcher's account permits reference to change if there is a shift in the ways in which a term's reference is fixed such that the set picked out prior to a given time differs from the set picked out after that time. However, in contrast with Devitt's notion of multiple grounding, Kitcher allows that one of the ways in which a term-token's reference may be fixed is by description. This accords with the considerations to be presented in the next section, and offers further possibilities of reference change. For example, Kitcher notes that the reference of a term might come to be fixed exclusively by a description which turns out to determine a different extension from the original extension of the term (1982, p. 340).

Finally, let us see whether the promise of referential stability offered by the causal theory of reference survives rejection of the exclusive reference-determining role of initial term-introductions. Of course, Devitt and Kitcher de-emphasize original introductions in order to permit reference change, so greater reference variance is to be expected. However, it does not follow from granting later applications of a term a reference-determining role that reference varies with change in the descriptions associated with a term by later speakers.

Speakers are still able to inherit the reference of a term from earlier use by a chain of communication even if its reference has been established in more than one way. Because of this there is no need for later speakers to associate the same descriptions with a term as earlier speakers. For, while Kitcher does grant descriptions a reference-fixing role, later speakers whose reference ultimately derives from such a reference-fixing description may not themselves be in a position to supply the description. Further, provided such speakers derive their reference from the same reference-fixing events, they may still disagree among themselves by describing the shared referent in incompatible ways.

Thus, even though reference can be altered by varying the objects to which terms are ostensively applied and by using new reference-determining descriptions, it does not depend on the descriptive content associated by speakers with a term at a given time. Moreover, given the manner in which a term may initially refer to a set of objects which shares members with the set to which it later refers, there may be extensional overlap even as a term's reference undergoes change. As a consequence, major change of theory involving profound variation of descriptive content need not result in radical discontinuity of reference.

2.6 A role for descriptions

In this section I will discuss two problems which require extension of the causal theory of reference to include a greater role for descriptions in reference determination. These two problems are the so-called "qua problem" and the problem of the reference of theoretical terms. The position which results from increasing the role of descriptions is a form of "causal descriptivism", since causal relations are supplemented by descriptions in order to fix reference.[31]

The qua problem arises with respect to the ostensive naming of observable objects and kinds.[32] It has to do with how an object or a kind is picked out qua a particular sort of object, or qua a particular kind. According to the causal theory of reference sketched in 2.4, a term may be attached to an object or kind at a naming ceremony in which an object or a sample of the kind is singled out by ostension. The object which is named in such a ceremony is determined as referent by perception of the object as part of the ostensive term-introduction. Alternatively, the extension of a kind term is determined as the set of objects (or the stuff) which bears an appropriate sameness relation to the sample ostensively picked out at the naming ceremony. In either case, perceptual contact with the object or sample-set constitutes a causal relation which purportedly determines reference.

But the question arises of which sort of object, or which kind, is named in such a ceremony. The qua problem, in short, is that the causal relation obtaining between the introducer of the term and the object or sample-set ostensively indicated does not itself suffice to identify either what sort of object is named or what kind the sample-set is a representative sample of. Causal contact alone does not determine reference. In naming a cat, say, merely perceiving the cat does not determine that it is a cat which is named. For, given only the perceptual relation, it may just as well be a time-slice of a cat, part of a cat, or even the natural kind cat. A determinate causal relation may obtain between the namer and the perceived entity, but such an entity instantiates numerous categories of relevance to the reference of the name.

Similar remarks apply to the naming of a kind. Mere perception of a sample-set does not determine which kind the sample is a sample of. Thus, consider an example of Sterelny's:

> Suppose I go to Mars and come across a catlike animal: I introduce the term 'schmat'. Schmats are animals bearing a certain relation to this paradigm local schmat I have just encountered. But what determines which relationship this is? For the schmat will be a member of many kinds. A non-exhaustive list would include: physical object, animate object, animate object of a certain biochemical kind, animate object with certain structural properties, schmats, schmats of a certain sex, schmats of a certain maturational state. (1983, pp. 120-1)

So far as the causal relation between the paradigm schmat and the introducer of the term is concerned, nothing privileges any of the kinds of which it is an instance over any of the others. Ostension by itself leaves it indeterminate whether schmats are singled out qua physical objects, qua animate object, or qua instance of any kind at all.

The lesson to be drawn from the qua problem is that the kind to which an object or sample belongs must be specified at the ostensive introduction of a term. To resolve the indeterminacy, some indication of the relevant category must accompany the act of ostension. While the category may be left implicit, the ostensive definition may state it explicitly. Thus, in naming a cat, one may say "Let's call that cat Tiger", making it explicit that a particular cat is named. Similarly, with Sterelny's schmats, the introductory definition of the term 'schmat' may specify that it is a particular species of animal which is named: e.g. "That species is a schmat". Thus, to explicitly specify the kind relevantly instanced by an ostended item, ostension may be supplemented by a categorial term (e.g. a sortal expression) stating the category to which the named object or sample belongs.[33]

Granting such a supplementary role to categorial expressions does not alter the basic causal nature of ostension. Introducing a term ostensively still involves a perceptual relation between language user and ostended object. Ostensive indeterminacy fails to obtain provided only that the ostensive definition contains a suitable categorial term (or if one is implicit). What then determines the extension of a natural kind term is the condition of belonging to the same kind as the ostended sample, where the relevant sort of kind is specified by a categorial term.[34]

The role so far accorded descriptions is minimal. The qua problem shows specification of category to be necessary if indeterminacy of ostensive reference is to be avoided. However, as we now turn to theoretical terms, it will emerge that descriptions have an even greater role to play. This problem has been dealt with by a number of authors.[35] I will discuss the views of Devitt and Enç, before adopting Nola's suggestion that descriptions of causal mechanism are needed to fix theoretical reference.

In my account of the causal theory of reference in section 2.4, I noted that terms which refer to unobservable entities do not permit ostensive introduction. Descriptions are needed to fix the reference of such terms. I took note as well of the suggestion of Kripke and Putnam that the necessary descriptions take the form of a causal description. In particular, they suggest that the reference of a term which refers to an unobservable entity may be fixed by a description specifying the observable effects for which the entity is causally responsible.

According to a suggestion of Devitt's, however, even such causal descriptions import too great a role for descriptions (1981, pp. 200-2). Devitt suggests that a description introducing a theoretical term may occur in referential use, so that the reference of the term is determined by causal relationship rather than description. Since the referent is unobservable the causal relation cannot be a perceptual one. But Devitt suggests it must be a relation "very like perception: quasi perception" (p. 201). He describes quasi perception as 'a relation consisting of an instrument "perceiving" the referent and our "reading" of the instrument: we are "perceiving the referent through the instrument"' (p. 201). In sum: Devitt's idea is that a speaker may achieve reference to an unobservable entity by means of a quasi perceptual link via the instrumentation used to measure or detect it.

Devitt fails to develop this suggestion in greater detail and, indeed, the prospects for development seem limited. The suggestion trades on an analogy between "perception" by an instrument and perception by a human observer, which is not implausible, especially for such instruments as microscopes and telescopes. But the analogy is weaker with instruments which do not simply extend human sense perception. The problem is that an instrument may be wrongly taken to measure entities of a certain kind, while measuring entities of a different kind, linking its users causally to entities to which they do not refer. But to allow reference to entities other than those measured by the instrument implies that causal links may be disregarded in favour of descriptive specifications in determining reference. Even were this not so, a version of the qua problem would have a similar consequence. For the purely causal link of the instrument to the entities measured cannot determine the category under which they are picked out.

Not only is Devitt's idea that descriptions need have no role in determining theoretical reference problematic, their actual role seems to be even greater than Kripke and Putnam allow. Consideration of the issue of reference failure shows that more is involved in fixing the reference of a theoretical term than specifying observable effects for which the entity to which it refers is responsible. For the idea that reference is fixed in such a manner is inadequate as an account of the reference of terms which fail to refer while purportedly referring to the entities which cause the specified effects.

The point is made by Enç (1976) in connection with the term 'phlogiston'. He notes that on Kripke's and Putnam's account scientists who used the term 'phlogiston' would have "meant to refer by that term to whatever it may be that was responsible for the phenomena they were trying to explain" (p. 267). But this would make 'phlogiston' refer to oxygen, for it would entail that:

> the reference of the term 'phlogiston' [was] fixed as the substance that is in fact responsible for calcination and combustion. And since oxygen is the substance that answers the description ... the phlogiston theorists would be said to have been talking about oxygen... [T]he Kripke-Putnam Thesis ... would [ ] prompt us to say that phlogiston theorists were in fact talking about oxygen and that they had some false beliefs about oxygen. (1976, pp. 267-8)

Enç takes this to be an unacceptable consequence, claiming to the contrary that:

> Lavoisier, in discovering oxygen, discovered that phlogiston does not exist ... phlogiston theorists ... were not talking about oxygen. They were not as a matter of fact talking about anything that exists. (1976, p. 268)

Because phlogiston theorists held quite specific beliefs about the nature of phlogiston and about its role in the causal processes underlying combustion and calcination, they failed to refer to oxygen. They did not mean to refer broadly to whatever it is that in fact causes combustion and calcination, but to a specific substance which produces those effects in a particular way.

Enç has a positive proposal designed to do justice to terms like 'phlogiston'. He claims scientists introduce a new term when their explanation of a set of phenomena leads them to believe a new kind of entity is responsible for the phenomena. According to Enç, in developing such an explanation scientists arrive at hypotheses about the properties of the responsible entities, and about the manner in which their causal action produces the phenomena. Typically, "the hypothesized properties and the conjectured explanatory mechanism bring along with them the suggestion that the object in question is of a specific kind" (pp. 270-1). If the kind is a new kind a new term may be introduced, and "the burden of reference for the term will be carried by the kind-constituting properties attributed to the object and by the explanatory mechanism developed by the theory" (p. 271).

The reason the burden falls on such properties and causal mechanism is that it is the belief that the explanation involves a specific kind of previously unknown entity that licenses the introduction of a new term.

> In other words, in introducing the term, the scientist is not just naming whatever it is that is responsible for such and such phenomena, he is rather naming a kind of object partially specified by the kind-constituting properties he believes the object to have and by the context in which the object plays its explanatory role. (1976, p. 271)

However, the motive for introducing a new term is not merely to refer to a specific kind, but rather to name a new one: "When the scientist introduces a term for a kind of object, linguistic parsimony urges that he believe that he is naming a new, hitherto unknown kind of object" (p. 271). Enç's emphasis on the motive of naming a new kind leads him to place more weight on properties characteristic of it as a kind than on its causal role: "the introduction of a new non-o[stensive]-term into our vocabulary implies that we believe we know first what kind of thing we are naming and second that the kind of thing we are naming is a new kind of thing" (p. 277).

However, Enç's added emphasis on the nature of the kind at the expense of causal mechanism seems misplaced. For while oxygen, not having the required causal role, is not phlogiston, it is less clear that an entity with the causal behaviour attributed to phlogiston need have any other properties to count as phlogiston. The point is that a term like 'phlogiston' is introduced in the context of an explanation for the specific purpose of referring to an entity which has a particular causal role. That is, the existence of such an entity is posited because an explanation of certain phenomena appeals to an entity whose agency brings about the phenomena in a specific manner. Given the explanatory purpose for which the entity is posited, properties of the entity unrelated to causal role are inessential, and it is unnecessary to specify them to determine reference. For this reason, the emphasis should be on causal role, not on kind-constitutive properties where Enç places it.

Nola, in his (1980b), agrees with Enç that the reference of theoretical terms is not determined by mere specification of effects, citing Enç's objection that 'phlogiston' would then refer to oxygen.[36] However, Nola disagrees with Enç on the role of kind-constitutive properties, and instead emphasizes causal mechanism:

> A scientist in observing phenomena O may hypothesize that one kind of non-observable entity is causally responsible in a particular way for O and he may begin to call the kind of entity by the name 'T'... In introducing the name 'T' to talk about T the scientist will also attribute causal powers to T such that it brings about phenomena O in a particular way. That is, the scientist will form beliefs about T of the form: all T have power P which in circumstances C cause O (for example, all phlogiston has the power to leave metal when the metal is heated thereby causing it to collapse into a powdery calx). (1980b, p. 524)

Specifying the causal powers whereby the action of an entity produces a given set of phenomena suffices to pick the entity out as referent. Moreover, difference in putative causal powers suffices to show that such entities as phlogiston and oxygen are distinct kinds; their unlike causal powers distinguish them even without specifying "precisely what kind each thing is beyond claiming, perhaps, that each is a substance" (1980b, p. 525).

The chief advantage of emphasizing causal mechanism is that less descriptive content is involved in determination of reference than if characteristic properties of the kind are specified. This increases the scope for stability of reference through variation of descriptive content. For while specification of causal powers provides enough information to discriminate between phlogiston and oxygen, it does not require so much descriptive detail as to make reference over-sensitive to adjustments of theory. This permits phlogiston theorists to vary widely among themselves over the specific properties and nature of phlogiston while continuing to use the term as a name for the substance whose release from burning bodies constitutes combustion.

Moreover, causal mechanism captures the key elements of a theoretical entity's explanatory role which are constitutive of the commitment on the part of a theory to an entity like phlogiston. For the phlogiston theory is committed to the existence of an entity which occupies the causal role specified by the phlogistic explanation of calcination and combustion. And for as long as phlogiston theorists employ such an explanation, they continue to putatively refer to an entity which acts in the described manner to produce the specified effects.

Finally, while reference variance is reduced by emphasis on causal mechanism over kind-related properties, it may nonetheless appear that such a reference-fixing role for descriptions risks a return to the referential instability of the description theory of reference. Here the reference-fixing role for descriptions joins with the fact that reference may be multiply fixed to suggest a greater possibility of reference change. Given the rejection of initial term-introductions, a causal role description used subsequently to fix the reference of a theoretical term may incur a change of reference if it is incapable of being satisfied conjointly with an earlier such description. More generally, if an earlier causal role description is rejected as mistaken and replaced by an incompatible new one, the reference will change through a changed conception of causal role. On the other hand, if there is no change in the causal role descriptions used to fix reference, or if subsequent ones are satisfiable along with earlier ones, reference may remain constant through variation of descriptive content which does not alter causal role.

Notes

1. As argued by Fine (1975).
2. See Feyerabend (1981f, p. 115), and Kuhn (1970b, p. 266) and (1976, p. 191). Feyerabend's concession and Kuhn's idea of "point-by-point" comparison are discussed in Chapter One; see sections 1.2 and 1.3 respectively.
3. Formally, Pa&~Qb, where P=Q and a=b.
4. English (1978, p. 63) distinguishes between "syntactic" accounts of contradiction which require uniform symbolization and "semantic" accounts on which synonymy suffices. She contrasts both of these with Scheffler's "coreferential" account, and notes that on Scheffler's account "contradictions no longer form a convenient class whose falsehood is necessary or can be seen without empirical research" (1978, p. 64). There has been little discussion of this issue in relation to incommensurability; but see Devitt (1979, p. 34) and Grandy (1983, pp. 19-20).
5. Ordinarily, it must be a token of the same lexicographic as well as the same semantic sentence-type.
6. The point that incommensurability is a relation resulting from semantical differences between theoretical sub-languages will be employed in Chapter 4 to defend the idea of an untranslatable language against the criticism of Davidson and Putnam.
7. More particularly, it is inconsistent to assert that an object or set of objects belongs to the contained set while denying that it belongs to the containing set. In his (1972, p. 253) Martin considers the case in which T_1 entails Ba, T_2 entails ~Ba, and "the referent of B has [ ] changed from T_1 to T_2". He notes that "as long as we know that the extension of 'B' in T_1 is a subset of the extension of 'B' in T_2 the two theories can be shown to be in conflict".
8. I will discuss Kitcher's views in more detail later. For now I note simply that his idea of the variant reference of term-tokens suggests a further means of referential comparison. In 2.5 his view on reference change will be dealt with, and his views on translation between theories figure centrally in sections 3.4-3.6.
9. Thus Kitcher's approach contrasts with Field's, for whom it is term-types rather than tokens which partially denote more than one entity. However, there appears to be no reason why the two approaches could not in principle be

combined; e.g. some tokens of a term might partially denote two or more kinds not referred to by other tokens of the term.

10. This conclusion is not restricted to terms which actually have reference. Terms may fail to refer and yet purport to refer to the same thing; in virtue of such sameness of purported reference, sentences containing the terms can come into conflict.
11. In fact, as a number of authors have commented, change of sense would tend to be accompanied by a change of a term's reference; change of sense not affecting reference is in certain respects accidental. See Devitt and Sterelny (1987, p. 182), Leplin (1979, p. 270), and Papineau (1979, p. 58).
12. Recall Feyerabend's specification of conceptual changes which result in incommensurability, discussed at the end of 1.2: "we shall diagnose a change of meaning either if a new theory entails that all concepts of the preceding theory have zero extension or if it introduces rules which cannot be interpreted as attributing specific properties to objects within already existing classes" (1981e, p. 98). Feyerabend's account of radical reference change will be dealt with in detail in 5.2.
13. E.g. "the physical referents of these Einsteinian concepts are by no means identical with those of the Newtonian concepts that bear the same name" (1970a, p. 102). For details, see the discussion in 1.3; Kuhn's original position on radical reference change is the topic of 5.3.
14. E.g. Kuhn notes against Scheffler that the "alloys were compounds before Dalton, mixtures after", hence "the reference of 'compound' ... changes" (1970b, 269). For general remarks on the transfer of subsets between such pre-existing categories, see Kuhn (1970b, p. 275) and (1981, p. 25).
15. The description theory of reference characterized here stems from the classic Fregean theory of reference. The appellation 'description theory of reference' appears to be due to Kripke (1972), who criticizes the classic account of reference as well as the later cluster theory of reference.
16. A fuller analysis of Kuhn's and Feyerabend's views on reference will be given in Chapter 5.
17. This reveals a disanalogy between the cases of interest to Kuhn and Feyerabend and standard examples of non-synonymous co-referential expressions: viz. a pair of co-extensive

expressions such as 'creature with a kidney' and 'creature with a heart' constitute adequate descriptions of the same thing. The point will be discussed in connection with translation in section 3.2. For a related point see Leplin (1979, p. 270).

18. The point that the description theory of reference leads to a thesis of referential incommensurability is a familiar one. See, for example, Devitt and Sterelny (1987, p. 182), Hacking (1983, p. 76), Newton-Smith (1981, p. 160), and Nola (1980a).
19. The real thrust of the story is that reference is independent of description. Thus the story may be modified to show that satisfaction of description is unnecessary. Just suppose that the description of water used on Earth fits XYZ but not H_2O; H_2O is still the referent of 'water' here on Earth.
20. Putnam expresses the point in terms of indexicality, noting that natural kind words like 'water' are indexed to the environment in which they are used: "words like 'water' have an unnoticed indexical component: 'water' is stuff that bears a certain similarity relation to the water around here" (1975b, p. 234).
21. Cf. Putnam, "'Water' on Twin Earth is not water, even if it satisfies the operational definition, because it does not bear $same_L$ [i.e. the relation same-liquid-as] to the local stuff that satisfies the operational definition" (1975b, p. 232).
22. Kripke (1972, pp. 325-6) and Putnam (1975a, p. 200).
23. The label "causal theory of reference" may suggest the view that causal relations are constitutive of reference. But this is a separate thesis independent of the view that causal relations play a large part in determining reference. Nor is it germane to the present purpose of showing that reference can withstand conceptual change to enter the issue of what reference in fact is.
24. Cf. Putnam, "concepts which are not strictly true of anything may yet refer to something, and concepts in different theories may refer to the same thing" (1975a, p. 197).
25. See Devitt (1979, pp. 40-1), (1981, pp. 191ff), and Devitt and Sterelny (1987, p. 72).
26. A 'grounding' is defined as a "perception (or quasi-perception) of an object that begins a reference determining causal-chain for a term" (Devitt and Sterelny, 1987, p. 253).
27. Admittedly, Devitt does extend the notion of grounding to include the use of instrumental links to objects which cannot

be directly observed, hence the mention of "quasi-perception" in the definition quoted in the preceding footnote (see also Devitt, 1981, p. 201). But while this is a promising suggestion for some terms, it would appear from considerations to be raised in the next section to have at best a limited application to only certain theoretical terms.

28. The basis for modifying Devitt's account is perhaps to be found in his concession that the reference of some theoretical terms is fixed by attributive description (1981, p. 202). This would allow a term whose reference is fixed in more than one way to have it fixed by description as well as by causal grounding. But Devitt defines grounding as a non-descriptive causal link, so building attributive descriptions into his account of reference change would not be to explain reference change by multiple grounding.
29. See Kitcher (1978, pp. 529-46) and (1983). The case will be discussed in greater detail in the following chapter, sections 3.5 and 3.6.
30. In his (1978) Kitcher suggests that co-reference of tokens permits context-sensitive translation, a position from which he retreats in his (1983). In section 3.6 I will argue against such contextual translation.
31. The position developed here has affinities with the "descriptive-causal" theory of reference to which Devitt and Sterelny are led by the qua problem (1987, pp. 72f). A related position is that reference is determined by descriptions couched largely in causal terms (see Lewis, 1984, pp. 223, 226).
32. The appellation "qua problem" seems to be due to Sterelny (1983, p. 120). For discussion of the problem see Papineau (1979, pp. 158ff) and Kroon (1985, pp. 145-7).
33. Arguably, this is already a feature of Putnam's original account of ostensive introduction of natural kind terms (1975b, p. 225). His example is "This liquid is called water", which defines 'water' as what bears the "same-liquid-as" relation to an ostended sample of water. Here the substance term 'liquid' removes the indeterminacy.
34. This appeal to categorial terms may seem to revoke a founding assumption of the causal theory of reference, viz. that reference may succeed even if the referent is misdescribed. But this is only apparent: categorial terms supplementing ostension may be mistaken. Even if use of the categorial term leads to misdescription, there may be a

way to replace or modify the description which still avoids the indeterminacy.

35. For a summary of the literature on the topic see Kroon (1985).
36. Nola (1980b, p. 522). Nola has an objection of his own. In brief: to fix reference by appeal to effects results in an ill-formed definition; for "objects do not stand in causal relations: events do" (1980b, p. 507). He suggests that specifying an object as a constitutive part of an event requires theoretical description of the object, which avoids unintended reference to whatever actually causes the effects. However, Kroon (1985, p. 155) shows that this will not do: modify the form of the definition of 'phlogiston' to range over events and it can still be made to pick out oxygen.

3 Translation failure between theories

3.1 Introduction

This chapter defends the idea of translation failure by expressions employed within one theory into the language of another theory.

Untranslatability has a central role in the incommensurability thesis. If a pair of theories is incommensurable, then the languages employed by the theories are partially or wholly untranslatable. In addition, translation and content comparison have a close connection according to the thesis. Since the content of theories expressed in untranslatable languages is inexpressible within a shared vocabulary, it appears not to be directly comparable.

This putative connection between untranslatability and incomparability is refuted by the referential approach to content comparison espoused in Chapter Two. Comparison of content requires only that expressions be related via reference, not that they have the same meaning. Without the dependence of comparison on translation, theories may be untranslatable yet comparable with respect to content by means of reference.

Advocates of the referential approach tend to avoid the issue of translation. Either from a realist's concern with truth and progress,[1] or out of a desire to overcome the problem of content comparison, they focus on relations of reference and ignore translation. But the referential approach to comparison does not remove the problem of translation. Expressions from one language may refer to the same things as expressions from another language without being translatable into it.

The idea of untranslatability has been severely criticized by Davidson and Putnam, who argue that the very idea of a language which cannot be translated is incoherent. If they are right then there is something fundamentally wrong with the idea of translation failure between theoretical languages.

The principal aim of this chapter is to argue for translation failure between theories within the framework of the approach to comparison sketched in the last chapter. In the next chapter the idea of untranslatability of theories will be defended against the arguments of Davidson and Putnam.

In outline this chapter is organized as follows. Section 3.2 is about the connection between translation and content comparison. Section 3.3 discusses constraints on translation. These are extended to untranslatability between theoretical languages in 3.4, and applied to particular cases in 3.5. Section 3.6 deals with a complication about contextual translation.

3.2 Translation, reference and comparison

Intuitively, it may seem that content comparison requires translation and that incommensurability entails incomparability. This intuition receives further support from the nature of comparison. Comparison requires the existence of something common, so comparable things must share features with respect to which they may be the same or different.

To be comparable, then, the content of theories must have something in common. Such content is expressed in language and must be compared in language. This suggests that content comparison requires a common vocabulary in which to contrast assertions and denials of a common set of propositions. But if theories are not intertranslatable, no common vocabulary exists, and no statement can be formulated on which there may be either agreement or dissent. So direct comparison of content is impossible.

This apparent connection between translation and comparison assumes a need for sameness of meaning. It presupposes that in order for statements to be comparable for content their constitutive expressions must have the same meaning as well as the same reference. But this is more than is needed for comparison. In section 2.2 we took note of Scheffler's point that in the absence of shared meanings content is comparable by means of shared reference. Provided that their constituent terms have common reference, statements divergent in meaning may be incompatible with respect to truth-conditions, and hence comparable for content.

Translation, however, presents a serious complication for this appeal to reference. As noted in section 2.3, it is not enough simply to point out that reference suffices in principle for the purposes of content comparison. It has to be shown for the cases in question that there actually is common reference. But to show that there is common reference requires the use of a theory of reference. The trouble is that, on the description theory of reference employed by Scheffler, untranslatable theories may fail to have any common reference.

According to the description theory of reference, non-synonymous expressions may co-refer since the same thing may be described in different ways. The problem with an approach such as Scheffler's is that it relies on a questionable analogy between the untranslatable terms of incommensurable theories and standard cases of non-synonymous co-referential expressions. Standard examples of the latter include expression pairs such as 'renate' and 'cordate', and 'Hesperus' and 'Phosphorus'. The characteristic feature of such pairs is that their descriptive content may be jointly satisfied in the sense that the same thing or things may instantiate the properties specified in their associated descriptions. For instance, all and only animals which satisfy the description 'creature with a kidney' instantiate the property specified by the description 'creature with a heart'.

The analogy between incommensurable concepts and non-synonymous co-referring expressions breaks down because the translation failure at issue is not due to mere difference of meaning. As was pointed out in section 2.3, Kuhn and Feyerabend discuss cases of conceptual disparity in which the descriptive content associated with the expressions of incommensurable theories is incompatible. For that reason the descriptive content of expressions which fail to be inter-translatable due to incommensurability cannot be jointly satisfied. According to the description theory of reference, therefore, such expressions do not co-refer. So if all of the expressions of one theory are untranslatable into another because of such conceptual disparity, the theories have no common reference, and their content is incomparable.

Thus the description theory can be seen to give further support to the view that untranslatable theories are incomparable. To fully disjoin comparison from translation requires another conception of reference. In effect, the causal theory of reference severs co-reference from translation. It allows that terms associated with jointly unsatisfiable descriptions can refer to the same things independently of the content of their defining descriptions. Such terms can function as co-designative expressions whose identical or intersecting extensions suffice for comparison, even if the incompatibility of their descriptive content precludes translation.

In this way, the referential approach adopted in this book extricates comparison from translation. The program for theory comparison which is based on a causal account of reference has no need to show that theories are intertranslatable.

3.3 Untranslatability

The special nature of incommensurability requires translation to be taken in a strict sense. While approximate translation may be standard practice in translating natural languages, the incommensurability thesis denies exact translation. For it claims that the terms of a theory have no semantical equivalents expressible in the language of a theory with which it is incommensurable. So translation must be taken in a strict sense as the formulation of expressions within a language which are semantic equivalents of expressions of another language. Translatability in this sense is a function of what can be said in a language. It depends on the ability to formulate semantic equivalents, and inability to translate reflects limits on what can be expressed in a language. This dependence is not affected by the fact that languages can be semantically enriched. Inability to formulate an equivalent without extending semantic resources constitutes failure of translation into the unmodified idiom.

Semantic equivalence is not word-to-word synonymy. A phrase can mean the same as a single word and be acceptable as its translation. Translatability depends on the ability of a language to define a term, not on whether it has a single word equivalent. Since phrases may translate single words, semantic and terminological change are distinct. The introduction of a novel term defined on the basis of extant semantic resources does not represent change in semantic resources. While translation fails if expression of an equivalent requires semantic enrichment, it may succeed if it merely requires terminological innovation.

I will take the term 'expression' as a generic term for referring expressions (terms, descriptions, etc.), and I will speak of pairs of expressions which are translations of one another as 'translational expressions'. Any discussion of untranslatability must employ some concept of translation. But, since failure rather than success of translation is at issue, a necessary condition of translation is all that is needed. For this purpose, it suffices to consider semantical properties of expressions which have to do with reference. I will begin by considering the relationship between the reference of translational expressions.

Since translational expressions are semantic equivalents the presumption may be that they have the same reference. Co-reference cannot, of course, be a sufficient condition of translation, since expressions with the same reference may differ widely in meaning. But it may seem to be a necessary condition, since terms with different denotations appear in sentences which vary in truth-value and make different assertions. The situation, however, is not this simple. It is unnecessary for a term to actually have reference in order for it to be translatable. Terms which fail to refer can still be translated (e.g. 'unicorn'/'licorne'). Thus the requirement of co-reference cannot be a requirement that a pair of terms succeed in referring to the same actual things, since terms which fail to refer would then be untranslatable.

Rather than co-reference in the actual world, what is required is sameness of possible reference. That is, translational expressions must have the same extension in all possible worlds.[2] This is because pairs of expressions whose reference diverges in different possible worlds are not semantically equivalent. Translational expressions which fail to refer in the actual world must be such that, for any possible world in which either has reference, they have the same reference. In general, whether or not translational expressions refer, they must have the same extension in all possible worlds. This is stronger than the requirement of actual co-reference, since it rules out pairs of expressions which contingently co-refer (e.g. 'renate'/'cordate').

Both actual and possible reference depend on the way reference is determined. The way reference is determined not only establishes reference in the actual world, it also determines reference for other possible worlds. So a pair of terms whose reference is determined in an identical way co-refers in any world in which either refers. Because of this, sameness of reference determination also constitutes a necessary condition of translation: translational expressions must have their reference determined in the same way. Indeed, we will see later in this section that sameness of reference determination is a stronger constraint on translation than sameness of possible reference. A principal aim of the theory of reference is to provide an account of reference determination. Since the way a term's reference is determined is a semantic property which must be preserved in translation, providing such an account sheds light on translation. For what semantically distinguishes non-equivalent co-referential expressions is the difference between the way in which their common referent is determined.

This is central to the traditional description theory of reference on which meaning is comprised of sense and reference, with reference determined by sense. Synonymity and translationality were both thought to entail sameness of sense and reference. Since sameness of sense meant sameness of reference determination, it followed from such a view that having reference determined in the same way was a requirement of translation.

The causal theory of reference does not rescind this requirement of translation. Rather than reject the connection between reference determination and translation, what the causal theory rejects is the traditional view of reference determination. Causal theorists have shown that the range of factors germane to reference includes physical and causal relations, rather than purely conceptual ones. The ambit of such arguments is restricted to the issue of how reference is determined. They do not bear upon the relation between reference determination and translation.

According to the causal-descriptivist approach adopted here, a pure description theory applies only to attributive descriptions. As opposed to a pure causal theory, however, causal-descriptivism does not assume there to be any cases in which reference is determined in a fully non-conceptual or non-descriptive manner. Let us consider several ways in which reference may be determined to see what translation must retain.

In the special case of terms whose reference is determined by attributive description, a set of properties is specified such that the referent is whatever it is that possesses those properties. For attributives, what determines reference is the satisfaction, by the members of some set, of the description specifying those properties.

The reference of a natural kind term may be fixed by ostension. The deictic component of the act of ostension does not alone suffice to determine reference to the kind: more than mere pointing is needed. Ostensive reference determination has to be taken in a wide sense which includes the act of pointing and the object pointed to. Additionally, since any ostended object belongs to a variety of kinds (e.g. liquid, beverage, chemical compound), deixis must be supplemented by a categorial term (e.g. 'liquid') which specifies the kind referred to. For example, the reference of 'water' might be fixed ostensively as 'the same liquid as the liquid in that glass', where a glass of water is indicated deictically. What determines reference in such cases is the condition of being the same stuff, as specified by the categorial term 'liquid', as the stuff present on the occasion of ostension.

The reference of natural kind terms may also be fixed by a description which contingently identifies the referent. The reference of 'water' may be fixed as the liquid which has the contingent property of flowing in rivers here on earth. Such a reference determination picks water out by a contingent property which water happens to have in the actual world. It does not make water the stuff, whatever it is, that flows in rivers: 'water' is not a word for anything that happens to run in a river. Rather, water is a liquid which just so happens to flow in rivers. It is the same kind of stuff as the liquid which, in present circumstances, flows in rivers. But things could change. The earth could freeze and water turn to ice. Liquid nitrogen might then run in the channels of present rivers, but would not in virtue of that be water. It is not the description that determines extension by what it is true of in different possible situations. Rather, as with ostension, what determines extension is the condition of being the same stuff, as specified by a categorial expression, as the stuff which has certain contingent properties here.

In translation the condition which determines reference must be isolated. To translate 'water' by an expression which refers to whatever flows in rivers would be to mistake the way its reference is fixed and to mistranslate it. To translate 'water' correctly an expression must be sought whose reference in every possible world is the same stuff as water, and which is specified by the same categorial term ('liquid' not 'chemical compound'). Of course, this does not hold only for rigid designators such as 'water'. Suppose we had a non-rigid term 'riverfluid' which refers to anything that flows in a river, not just water. The translation of 'riverfluid' would have to mirror its reference in possible worlds, including ones in which riverfluid is something other than water.

It might seem that the demand of identity of reference determination adds nothing in point of stringency to the requirement of sameness of possible reference. However, there are pairs of terms which co-refer in all possible worlds, but whose reference is determined in different ways. Such terms cannot be distinguished by appeal to possible reference. Yet the difference in the way their reference is determined constitutes a difference of semantic content. So the demand of sameness of reference determination is a stronger constraint on translation.

For example, consider the following two expressions which denote the number ten: '8+2', '5x2'. These expressions denote ten in all possible worlds, but ten is determined as their referent by different means. The first uses the operation of adding eight to two, while the second multiplies five by two. Another example is that of 'cube' (i.e. a solid bounded by squares) and 'regular polyhedron with six square faces', which have the same extension in every possible world but specify it via non-equivalent descriptions.[3] A less intuitive example involves pairs of natural kind terms which refer to the same thing in all possible worlds, but which have their reference determined by different conditions. Consider, for example, the co-referential pairs 'water'/'H_2O', 'salt'/'NaCl', where the second members of each pair, unlike the first, have their reference determined in a way which depends on a particular system of chemical classification.

The class of terms in question is one in which different means of determining reference persistently determine the same referent in all possible situations. The relevant difference in the determination of their reference is conceptual rather than ostensive. With '8+2' and '5x2', mathematical operations are specified which constitute different ways of describing the number ten: viz. 'the sum of eight and two', 'the product of five and two'. With natural kind terms such as 'water' and 'H_2O', the difference lies in the categorial term which specifies the kind: 'same liquid as' versus 'same chemical compound as'.

Such difference of reference determination induces difference in semantic content. Different information is conveyed about the referent by different means of determining reference. To say that ten is obtained by multiplying five by two conveys different information about ten from saying that it is obtained by adding eight to two. That a cube is a solid whose faces are square is a distinct item of information from the fact that it is a particular sort of regular polyhedron. The information conveyed by saying that table salt is used to season food and is found in sea water is different in kind from information about its chemical make-up that is conveyed by referring to the same stuff under the heading of 'sodium chloride'.

To take the case of water, a difference of content is registered if it is referred to as 'H_2O' rather than as 'water'. The information conveyed by the category terms involved in fixing their reference is of a different kind. Referring to water as 'H_2O' brings to bear information about the chemical composition of water, that it is a compound of hydrogen and oxygen rather than an element. Such information depends in turn on the modern theory of chemical elements and composition. The categorial term 'chemical compound' used to fix the reference of 'H_2O' is conceptually distinct from categorials such as 'liquid' used for 'water'; it expresses different information about water.

Since such conceptual apparatus is built into the way reference is fixed, it follows that the conditions used to determine reference depend on an epistemic and theoretical background. Different ways of determining reference are therefore an index of difference in theoretical or epistemic content. To describe an object by means of a term which has its reference fixed in one way therefore conveys information which may well differ from that conveyed by using another vocabulary. Consequently, for the purposes of translation, where the aim is to formulate semantically equivalent expressions, it is necessary to use expressions whose reference is determined in the same way. Thus one way of showing that an expression is untranslatable into a theoretical or conceptual framework is to show that, within that framework, its reference cannot be fixed in the manner required.

3.4 Untranslatability between theories

This section extends the discussion to untranslatability between theories. The following criterion of untranslatability may be formulated on the basis of the preceding discussion of reference determination:

> A term from one theory is untranslatable into the language of another theory if no expression whose reference is determined in the same way is formulable in that language.

Defined for expressions instead of terms, the criterion allows for the possibility of translation where no word-for-word equivalent exists.

It is a criterion of untranslatability in the sense that it sets a sufficient condition of failure of translation. It is the converse of the necessary condition of translation just argued for, from which it arises by taking failure to meet the condition as sufficient for failure of translation.

The criterion needs to be refined in a crucial respect. Terms employed in scientific theories may be associated with multiple determinants of reference. Thus translations may need to preserve several reference determinations.

As we saw in the preceding chapter, Kitcher argues that different tokens of the same scientific term-type may have their reference fixed in different ways:

> When we look at the language in use among scientists at a particular time, we may find that for some important expression types there is a variety of ways in which the reference of tokens of those types can be fixed, and that the varied employment of tokens of these types presupposes connections that later scientists will reject. So, from the perspective of the scientific language in use at later times, the former usage of the key terms will be mistaken, and there will be no term in the later language which is used in the same variety of ways as the old expressions. (1983, p. 694)

Kitcher's basic point is that a term's reference may be fixed in a number of ways because it may be applied in different contexts. Members of the same kind may be present in or thought to be present in a variety of situations. They may also be described in different ways. So there may be alternative ways in which the reference of tokens of a term for that kind may be fixed.

The point emerges clearly in connection with terms whose use includes both an ostensive and a descriptive component. Within a theory a term may be applied directly to observed samples as well as being defined by a description of the purported nature of the kind to which the samples belong. Such diverse use seems unified because it is assumed that the kind to which the ostended samples belong is identical with the one specified by the description. But such a unifying assumption may, of course, be false. In that case the term's extension will be heterogeneous, since non-identical kinds will be picked out by the two reference determinations.

Natural kind terms such as 'water' exhibit the same diversity. We can give a variety of ostensions and descriptions of water which seem to pick out the same natural kind. Because we believe that they pick out the same kind, the existence of multiple means of fixing reference only suggests to us that there are a number of different contexts in which the referent of the term is found.

Evolution of the use of theoretical terms reveals the same phenomenon. For example, a term may be introduced to refer to an entity which is conjectured to perform a particular causal role. Subsequent research may then lead to a more complete specification of the properties and behaviour of the postulated entity. Later still the entity may be isolated experimentally, and the term applied to it in a more direct way. The application of such a term goes through a development in the course of which a number of different ways of fixing its reference accrue to it.

However, Kitcher's point need not entice us into an overly fragmentary picture of semantic content. It is not as if such terms may splinter into their tokens and translate severally. To translate token for token would result in loss of content. It would remove the presupposition of unity of reference underlying the diversity of reference determination. The point of retaining the same term even though its reference is fixed in different ways is that the set variously so picked out is presumed to be the same one. The set is taken to be unified by the homogeneity of the natural kind of which its members are constitutive. To remove the implication that the set constitutes a single natural kind by breaking down the connection between tokens alters the information conveyed by use of the term and is therefore a failure of translation. I will discuss this point in 3.6.

To accommodate Kitcher's point the criterion of untranslatability must be adjusted. It must take into account that scientific expression types can be associated with more than one way of determining reference. So we may amend the criterion as follows:

> A term-type is untranslatable into the language of a theory if no expression whose reference is determined in the same set of ways is formulable in that theory.

So adjusted, the criterion allows for the possibility that terms may have their reference fixed in a number of ways, although of course they need not do so. Notice also that the criterion applies to term-types, which prevents terms whose reference is multiply determined from being translated token by token.

The criterion may seem to be open to the following objection. It prevents a term whose use is extended by having its reference fixed in a novel way from being translated homophonically. For the reference of such a term would no longer be fixed in exactly the same set of ways. But, in the first place, this is in some cases the right thing to say. The way a term's reference is fixed may be altered in such a way that its semantic self-identity breaks down.

But such a change of meaning does not of course entail untranslatability. Even if a term's meaning changes between theories it may still be translated from one theory into the other in some other way. More importantly, the criterion does not in fact apply to such cases. It says nothing at all about the univocality of particular terms. Rather, the criterion is designed to deal with inability to translate a term into the language of a theory taken as a whole, not the synonymity of individual terms.

One further aspect of the criterion requires comment. What does inability to formulate an expression with the right semantic features amount to? What is involved in not being able to determine reference in a particular way?

There seem to be two possibilities. It either involves inability to attribute reference or a limit on the way reference can be determined.

Let us consider the first. It might seem intuitive that being unable to assign a referent to an expression of another language entails that its reference cannot be fixed. Suppose that in a language L a term t is taken to have a referent. We want to translate t into another language L*. What if no referent is attributed to t from the point of view of L*? As far as L* is concerned t is an empty term. This could occur in either of two ways. First, in L* reference can be fixed in just the same way that the reference of t is fixed in L, but according to L* nothing is picked out by that way of fixing reference. (For example, French speakers assume there are licornes, English speakers that there are no unicorns, and the reference of 'licorne' and 'unicorn' is fixed in the same way.) But in such a situation there is no reason to deny that t is translatable into L*. ('Licorne' could still be the translation of 'unicorn'.) The fact that no referent is assigned to t is beside the point. If it is possible to specify in L* the condition which determines the reference of t, it does not matter for translation if t is taken to be empty.

The second way is for t to have no referent when viewed from the vantage of L*, and for it to be impossible to determine reference in L* in the way that the reference of t is fixed in L. In such a case t is untranslatable into L* simply because reference cannot be determined in the required way. Again, it makes no difference whether t is taken to be empty. Even if a referent were attributed to t by L*, it would be untranslatable.

This returns us to the other possibility: that there are specific limits on the way reference can be determined in a language. But why should it be impossible to determine reference in one language in a manner which is possible in another?

In the case of scientific theories such limits are due to closure of the logical content of a set of theoretical principles. Assertions which are incompatible with a set of theoretical principles cannot be derived from those principles. Expressions formulable within the language of a theory must be defined on the basis of the principles of that theory.

Intuitively, a concept is indefinable within a theory if the theory does not permit it to be formulated. The basic principles of a theory consist of existence postulates describing the entities which populate its domain and laws governing their behaviour. Concepts formed on the basis of one set of principles may be incapable of formulation in a theory which rejects those principles.

Feyerabend has consistently stressed that concepts of some rival theories cannot be interdefined because their laws are incompatible. He has argued that "the classical, or absolute idea of mass, or of distance, cannot be defined within [general relativity]" (1981f, p. 115).[4] He argues that "the concept of impetus, as fixed by the usage established in the impetus theory, cannot be defined in a reasonable way within Newton's theory" (1981d, p. 66). Generally stated, Feyerabend's idea is that "the conditions of concept formation in one theory forbid the formation of the basic concepts of the other" (1978, p. 68, fn. 118).

Until recently, Kuhn did not make the point explicitly. He came closest to doing so in his original discussion of the non-derivability of Newtonian from Einsteinian laws (1970a, pp.

101-2). But there he emphasizes difference of meaning instead of indefinabililty. Of course, that argument relies tacitly on indefinability, since mere difference of meaning is insufficient: to show non-derivability it has to be shown that there are no definable equivalents.

Kuhn now explicitly speaks of "the impossibility of defining the terms of one theory on the basis of the terms of the other" (1983, p. 684; cf. p. 669). On his present account, the central concepts of a theory are interdefined relative to the basic laws of the theory, and as a result are untranslatable into a theory with different laws: e.g. "Newtonian 'force' and 'mass' are not translatable into the language of a physical theory ... in which Newton's version of the Second Law does not apply" (1983, p. 677).

So, in general, indefinability is due to incompatibility of theoretical principles. Applying the point to our approach, inability to determine reference in the required way must therefore be due to a prohibition on the part of a theory against that way of determining reference. In terms of our criterion, this means that translation may fail in two basic ways. Either a theory prohibits a particular way of fixing reference or it rejects a connection purported to obtain between different means of fixing reference. The two may, of course, occur in conjunction.

As for the first, a particular way of fixing the reference of a term may be unavailable in or may be rejected by a theory. This would make it impossible to define an equivalent expression within that theory by formulating the requisite condition of reference. An obvious case is the description of a type of entity, e.g. waves or phlogiston, whose existence is dispensed with or denied altogether in the framework of some theory. In such a situation, the description of the causal role needed to fix reference to such entities could not be formulated in the opposing theory.

The second way for translation to break down would be the rejection by a theory of the conjoint use of two or more ways of determining the reference of a term. An example would be disagreement between theories on the existence or identity of some natural kind. Suppose a theory ostensively attaches a term to several presumed samples of the same substance, which are obtained by different procedures. Another theory may deny that the samples of stuff so obtained belong to the same single kind of substance. As a result, the latter theory would be unable to define the term, since it denies the connection between the separate means of fixing reference.

To judge superficially from the writings of Kuhn and Feyerabend, there are two major causes of translation failure between theories: ontological incompatibility and difference of classificational system. Feyerabend usually speaks of changes of ontology. Thus, in his original argument for incommensurability Feyerabend claimed that in the transition between "general theories" T and T' there is "... a replacement of the ontology ... of T' by the ontology ... of T, and a corresponding change of the meanings of the descriptive elements of the formalism of T'" (1981d, p. 45). Incommensurable theories possess incompatible ontologies, which creates conceptual disparity and hence untranslatability.

> Changes of ontology ... are often accompanied by conceptual changes. The discovery that certain entities do not exist may prompt the scientist to re-describe the events, processes, observations which were thought to be manifestations of them and which were therefore described in terms assuming their existence ... [W]hen the faulty ontology is comprehensive ... every description inside the domain must be changed and must be replaced by a different statement (or by no statement at all). (1975, p. 275)

By contrast, Kuhn tends to speak of change of classification or taxonomy. He favours such images as this: "languages cut up the world in different ways" (1970b, p. 268), and "languages impose different structures on the world" (1983, p. 682). And on his view translation fails because the languages of different theories possess incompatible categorial schemes, between which individual categories are not interchangeable in a piecemeal fashion.

But despite this difference in favoured terminology, the difference is superficial and at most a matter of degree. For classificational systems invoke categories which are purported to exist, while ontologies invoke entities and categories in terms of which they classify. Ontological and classificational change are two sides of the same coin.

To be sure, the distinction can be made in principle. For a change in the categories into which a domain is divided could occur without any change in the sort of entities supposed to populate the domain, and vice versa. Admittedly, the two cases we are about to discuss are most readily thought of as cases of ontological incompatibility. But we will see in 3.6 that when the question of translation is pursued the issue of ontological incompatibility soon merges with that of classificational difference.

In general, the ontology of a theory is what exists according to a theory. More specifically, a theoretical ontology is a set of objects, kinds, magnitudes and processes. A theory postulates their existence and describes their purported behaviour in order to explain the phenomena in its area of investigation. An ontology is incompatible with another if it either rejects or allows no place for entities purported to exist by the other ontology.

It is not a sign of incompatibility of ontology if a theory does not have a word for an entity referred to by another theory. The real test is whether an expression with the same way of fixing that entity as referent is formulable in the theory. If such a way of fixing reference would violate principles of the theory, there is ontological conflict, and translation is impossible. Even if, as could well happen, the theory were able to refer to the entity using some other means of fixing reference, translation would still fail; for mere co-reference is insufficient.

3.5 Two cases of theory untranslatability

In this section I will apply the criterion of untranslatability to two cases. The first is Feyerabend's example of impetus. The second case is phlogiston versus oxygen chemistry, which has been the subject of a debate between Kuhn and Kitcher.

Feyerabend's impetus example is an important instance of ontological incompatibility (1981d, pp. 62-9). The existence of impetus is incompatible with the basic principles of Newtonian physics. The idea of impetus was formed within the system of thought of medieval Aristotelianism. According to the Aristotelian theory of motion, all motion requires the action of a continuous cause to sustain it.[5] Impetus was conceived as a force imparted to, or "impressed" in, a body by external action upon it. It was meant to be a force actually inside a projectile ("an inner moving force", Feyerabend says), which continues to propel it after physical contact between the body and its external influence is broken off.

The idea of impetus as a force which acts in a constant manner on all projectile motions conflicts with the Newtonian idea of inertial motion. From a Newtonian standpoint, a body in an inertial state — i.e. either at rest or in uniform rectilinear motion — is not under the influence of any cause. Thus no force acts upon a body in a state of inertial motion to cause its motion to continue. From such a point of view there is no such thing as impetus.

In order to translate 'impetus' into the language of Newtonian physics it must be possible to formulate an expression whose reference is determined in the manner in which the reference of 'impetus' is fixed in the impetus theory. But to do this, a theoretical description of the purported causal role of impetus must be formulated. Any description of the causal role of impetus must state that it is a kind of force and that it acts upon all projectiles, including projectiles in a state of inertial motion. But, as such, the definition of 'impetus' describes a specific causal agency whose existence is denied by Newton's theory. It is not therefore possible to formulate the requisite reference-fixing description on the basis of the principles of Newton's theory. As a consequence, 'impetus' cannot be translated into its language.

The reference of some tokens of 'impetus' could also be determined by ostension. It was not just a concept with a theoretical definition, for it was also applied directly to observed physical phenomena. Its reference could therefore have been fixed in two ways: by theoretical description and by direct ostension.

As it happens, the application of 'impetus' to empirical phenomena coincides with that of the Newtonian concept of momentum. Impetus and momentum admit of the same procedure of measurement, which yields an identical quantitative result for each.[6] Since such empirical application is not the only way that the reference of 'impetus' is fixed, joint ostensive application does not justify translating it as 'momentum'. Even though some tokens of the two terms are applied to the same motions, it is assumed on the part of the impetus theory that the motion to which 'impetus' is applied is a motion actively sustained by the sort of internal force which the theoretical definition of impetus describes. But since classical mechanics denies the existence of such a force, it denies the connection thought to obtain by impetus theorists between the two ways of fixing the reference of the terms. It denies that a description of an internal cause and an ostension of a bodily motion pick out the same thing.

Moreover, it should not even be assumed that the ostensions made in the empirical application of 'impetus' and 'momentum' determine reference in the same way. For the ostensive definition of 'impetus' must employ a sortal which specifies impetus as an internal cause, whereas that for 'momentum' must specify it as a quantity or measure of motion. Though they are applied in the same empirical situation, their ostensive reference determinations are in fact distinct.

It might be objected that under a more general sortal (e.g. 'physical magnitude') the ostensions are identical. But such an objection must face a version of the qua problem, viz. that ostension under such a general sortal succeeds only in fixing reference in a general way. Such ostensive reference is ambiguous between impetus and momentum, and needs further precision to fix reference to one or the other specifically.

I will turn now to the second case, that of phlogistic versus oxygen chemistry.[7] The main point of contention between the two chemical theories concerned the nature of combustion and calcination. Controversy flared up over the existence of a "principle of combustibility", known as 'phlogiston', which was thought to have the main causal role in combustion and the production of calxes. Proponents of the oxygen theory sought to explain those same chemical processes by means of an opposed causal mechanism, based on an ontology which dispensed with phlogiston.

In the background of this debate was the rise of pneumatic chemistry, which ultimately undermined the phlogiston theory. During the third quarter of the eighteenth century the number of gases known to chemists increased greatly. But pneumatic chemists did not see it in that way. It is anachronistic to say that "fixed air", "inflammable air", "dephlogisticated air", and the other "airs" discovered by Black, Cavendish, and Priestley were gases.

For the concept of a gas as a distinct chemical substance in a particular physical state came afterwards. "Common" or atmospheric air was not at that time thought to be made up of different chemical elements in a gaseous state. Air was taken to be a distinct substance in its own right: not a mixture of other elements, but itself an element in its own natural state. Conceiving air to be an element, phlogistic chemists saw the new gases as "airs" rather than gases. They took them to be so many different varieties of air: each air a modification of the air, with distinctive properties due to the presence of phlogiston and various impurities.

Phlogistic chemists held that combustion is the release of phlogiston. When a flammable substance such as wood burns it emits phlogiston into the atmosphere. Phlogiston is a chemical principle. A principle is a basic constituent of a body, whose presence in a body gives it certain properties, e.g. combustibility, fluidity. The presence of phlogiston gives a substance the property of inflammability, and its emission causes the loss of that property. Air has a limited capacity to absorb phlogiston, so combustion in an enclosed space stops when phlogiston saturates the air, which produces "phlogisticated air".

Calcination is analogous to combustion. When heat is applied to a metal it decomposes into its calx and releases its contained phlogiston. Thus a metal is more complex than its calx, since it contains phlogiston and calxes do not. Since loss of phlogiston causes a metal to lose its characteristic metallic properties, it is the presence of phlogiston that gives metals such properties. Calxes re-convert to metal when heated together with a source of phlogiston, such as charcoal. The phlogiston combines with the calx to produce the metal.

The contrasting picture yielded by the oxygen theory wholly inverts these processes. Instead of being in the burned or calcined substance beforehand, oxygen is in the air. So combustion and calcination are processes of combination rather than decomposition. Moreover, the oxygen theory rejects the idea that the various "airs" are modifications of elemental air. It conceives of them instead as distinct chemical elements and compounds which combine in a gaseous state to form atmospheric air. (Not until Dalton is the atmosphere thought of as a mixture rather than a compound.)

In combustion, oxygen from the atmosphere is taken on by the burning substance, with which it combines. When heated, a metal oxidizes by combining with oxygen. And an oxide re-converts to the metallic state by releasing the oxygen it has taken on in oxidation. Thus metals are elemental while their oxides are compound.

The oxygen and phlogiston theories, therefore, give completely different accounts of the processes of combustion and calcination. On the latter, a substance breaks down, emitting a contained principle into the atmosphere. While on the former, a gas is removed from the atmosphere, which combines with the substance.

But even though the structures of these processes are in opposition, the causal function of phlogiston is not entirely eliminated from the oxygen system. For the theory of oxidation was combined with a theory of the gaseous state. Lavoisier held that chemical substances become gaseous by combining with the "matter of heat", which he called 'caloric'. Thus oxygen gas consists of oxygen ("base of oxygen") combined with caloric. When oxygen is consumed by a burning substance or in calcination, the caloric disengages from the oxygen and escapes into the atmosphere while the oxygen combines with the substance. So according to the oxygen theory something does escape into the atmosphere in combustion and calcination. But the analogy between phlogiston and caloric goes no further than that. The released caloric is not contained beforehand in the oxidized substance, and phlogiston does not combine with chemical substances to put them into a gaseous state.

At the level of ontology, oxygen and phlogiston theory are incompatible. They are committed to different entities: phlogiston and various airs versus oxygen and other gases. They conceive common entities differently: elemental air and its varieties versus air as a a combination of elemental gases; compound metals and simple calxes versus elemental metals and compound oxides. And they conceive the processes of combustion and calcination in opposite terms: decomposition and emission of phlogiston versus consumption of oxygen.

Can 'phlogiston' be translated into the language of the oxygen theory? Ignoring the complexity of caloric momentarily, it is tempting to answer as follows. The reference of 'phlogiston' is fixed in the phlogiston theory by a description of a causal role, viz. phlogiston is that stuff, whatever it is, which calcined metal or burning matter emit. On the oxygen theory, something is taken on, nothing is emitted. So nothing fulfills that causal role, and no process satisfies that description. The necessary reference-fixing description cannot be formulated in the oxygen theory, and as a result 'phlogiston' cannot be translated into it.

The drawback with this argument is easy to see, for something is given off in oxidation, viz. caloric. So it is not counter to the oxygen theory to describe something let off in combustion. 'Phlogiston' cannot on that account be considered untranslatable.

There is, however, more to phlogiston than emission in the process of combustion. The emission of phlogiston is the very process of combustion, and its prior presence in the burned substance is necessary for combustion to occur. But on the oxygen theory there is no such stuff. For combustion is a process of consumption rather than elimination. It is the combination with something taken on from the air, not the release of something previously contained in the body. And what is required for combustion is the presence of oxygen in the atmosphere, not the presence in the body of something waiting to be released.

'Phlogiston' cannot be translated into the oxygen theory. It is not that nothing is given off in combustion. The point, rather, is that nothing contained in a substance beforehand disengages and escapes from the substance as the very process of combustion itself. It is precisely a reference-fixing description of the latter sort which cannot be formulated on the basis of the oxygen theory of combustion. Even if something is let off, combustion is not itself the process of emission, but the process of combination with something taken on.

It may seem that a phlogistic term with direct empirical application is more readily translated. In this context Kitcher has discussed Priestley's discovery of oxygen.[8] Priestley called oxygen 'dephlogisticated air'. Lavoisier, who learned from him how to produce it, later introduced the term 'oxygen' for it.

Priestley heated red calx of mercury with a burning lens, obtaining mercury and a new "air" with striking properties. Most notably, the air was good to breathe and supported combustion better than common air.[9] Since calx contains no phlogiston and metal does, the mercury calx must have taken on phlogiston in converting to mercury. Because the new air was obtained during a reaction in which phlogiston was consumed, Priestley reasoned that it must be air from which phlogiston had been removed. That would explain why the air supported combustion so readily, since air unsaturated by phlogiston would have room to take on emitted phlogiston. Because he supposed phlogiston had been removed from it, Priestley named the new air 'dephlogisticated air'.

Lavoisier interpreted the reaction differently. For him the new air was oxygen gas. It was released from the mercury oxide when the oxide was heated. And it supported combustion so well because it was itself the gas whose presence is needed for combustion to take place.

In a rough and ready way perhaps 'oxygen' and 'dephlogisticated air' are semantic equivalents. For once Priestley's use of the expression was established it became common use among phlogistic chemists to refer to the new air as 'dephlogisticated air'. Oxygen chemists could simply take 'dephlogisticated air' as the phlogistic name for oxygen.

But in the strict sense of equivalence required for intertranslatability as defined here the two expressions are not equivalent. The fact that they co-refer is not sufficient for translation, for their reference must be fixed in the same way. It will now be argued that 'dephlogisticated air' cannot be translated into the oxygen theory, from which it follows a fortiori that it cannot be translated as 'oxygen'.

The reference of 'dephlogisticated air' can be fixed in two ways. The first is by theoretical definition. According to the phlogiston theory, dephlogisticated air is air from which phlogiston has been removed, which soaks up phlogiston emitted in combustion. But from the standpoint of the oxygen theory there is no such thing as phlogiston which could be removed from the air. So 'dephlogisticated air' cannot even be defined in terms of the oxygen theory, which is to say that such a reference-fixing description cannot be formulated within it.

Secondly, the referent of the expression can be fixed by ostension. Priestley referred to the air obtained in the experiment with mercury calx as 'dephlogisticated air'. The fact that he obtained an air from the calx which he was able to manipulate experimentally implies the existence of an ability to ground the term in the substance ostensively.

Since Lavoisier had the same ability, 'oxygen' and 'dephlogisticated air' were both ostensively linked to the same substance. But even though both expressions are grounded in oxygen, the ostensions are not equivalent. The categorial specification required to narrow down which sort of substance is ostended must vary between the ostensions. For Priestley, the ostended substance was an air, a modification of elemental air. Whereas, for Lavoisier, the oxygen was a gas, a chemical element in a state of expansion, not a modification of elemental air. Not only are the ostensions non-equivalent, but Priestley's has content (owing to the sortal 'air') which is incompatible with the oxygen theory.

Even if it were allowed that the ostensions were the same, the term would still not be translatable into the oxygen theory. Our criterion requires that all of the ways in which the reference of a term-type is determined must be reproduced if it is to be translated. The reference of 'dephlogisticated air' is determined by theoretical definition and by ostension. According to the oxygen theory, the substance ostended by Priestley is not the substance left over when phlogiston is removed from the air. That is, the oxygen theory denies the connection purported to obtain by the phlogiston theory between the ostended substance and the theoretically defined stuff. So, even if the ostensions were the same, no expression having its reference jointly determined by both these means is formulable in the oxygen theory.

3.6 Against contextual translation

The overlapping application of 'oxygen' and 'dephlogisticated air' raises the possibility of contextual translation. Perhaps occurrences in which 'dephlogisticated air' is directly applied to oxygen may be equated with 'oxygen'. Other occurrences could be left untranslated or given a loose gloss such as 'air from which phlogiston has been removed'.

In his (1978) Kitcher held that 'dephlogisticated air' was translatable on a token-by-token basis. He proposed that the term could be translated in a context-sensitive manner by specifying the referents of its various tokens.[10] Thus tokens whose reference is fixed by ostension translate one way, tokens with reference fixed by theoretical definition translate in another.

But the idea that such expressions may fragment in translation must be rejected. (The point is not that translation must be word-for-word, but that it must be uniform.) The reason why contextual translation is unacceptable for expressions of this type is independent of the arguments just given that neither token of 'dephlogisticated air' is translatable into the oxygen theory.

Translation replaces expressions with semantic equivalents in another language. But to translate 'dephlogisticated air' as 'oxygen' at one point and as 'the air from which phlogiston has been removed' at another is not to replace it with an equivalent.[11] Rather, it is to replace it with two semantically distinct expressions. But expressions thus distinct are not equivalent to the original, for that would imply the equivalence of non-equivalents. Thus, such contextual translation is not translation.

This is not to rule out the idea of a context-sensitive translation in principle. Genuine ambiguity demands such sensitivity. Different tokens of an ambiguous term-type are semantically distinct.[12] They can be translated in a context-sensitive manner by semantically distinct terms.

But if Kitcher's point is that 'dephlogisticated air' is ambiguous, then its semantic content must divide into distinct components. For compare it with a term like 'bank'. Translated into French it splits in two, coming out as 'rive' and 'banque'. Different tokens of English 'bank' translate differently into French. But such tokens of 'bank' have distinct content in English, which are not semantically linked in any way. On some occasions, 'bank' means 'financial institution', on others 'side of a river'. The only thing the two uses have in common is the inscription 'bank'.

Nothing of the sort holds with 'dephlogisticated air'. Priestley applies the term to the newly discovered air because he thinks he has isolated dephlogisticated air. So far as he can tell, the air is a sample of a substance which it is possible to describe on the basis of the phlogiston theory. That is, he believes that the stuff let off by the mercury calx is the very stuff that is described in the phlogiston theory as air from which the phlogiston has been removed.

The fact that the reference of tokens of the term varies does not affect the issue. Priestley was under the impression that the reference of his tokens of the term was uniform. He did not knowingly use it to refer differently. Nor does difference in the way reference is fixed imply ambiguity. We noted at the beginning of 3.4 that tokens of the same term-type which are applied in different situations may well have their reference fixed in different ways.[13]

What determines that the term is unambiguous is that all of its occurrences were meant to apply to a single kind of substance: viz. air with phlogiston removed. In each of its separate uses it was thought to refer to the substance in general or to particular samples of the substance. Its diverse applications are unified by intended denotation of a single substance, since throughout those applications the concept of dephlogisticated air as defined in the phlogiston theory remains constant.

Contextual translation loses sight of such semantic connections between tokens. It treats tokens as semantically independent and obscures their intended uniformity. In so doing it alters the content of the tokens because it obliterates their semantic relation to the classificatory system that defines them. The whole point of using terms like 'impetus' and 'dephlogisticated air' on different occasions is that the term-type is presumed to refer to a single kind throughout the various applications of its tokens. What is picked out in different ways and situations is meant to belong to a kind which is quantified over within the ontology of the theory which defines the term.

Thus contextual translation loses content necessary to translation. The way to preserve it is to translate such terms as types and to insure that their translation refers to the relevant theoretically described kind. To refer to the same kind it is not sufficient that an expression merely have the same extension as another, for the same set can belong to more than one kind.

Rather, their common extension must be specified with the aid of equivalent categorial expressions which indicate what sort of kind the set belongs to. The constraint that the kind referred to by a term must be preserved in translation accords well with our criterion of untranslatability. The criterion guarantees that no term will be translated by an expression which fails to represent the same kind.

Kuhn has objected to Kitcher's "context-dependent strategy" in a similar manner.

> 'Phlogiston' would then sometimes be rendered as 'substance released from burning bodies', sometimes as 'metallizing principle', and sometimes by still other locutions ... Use of a single word 'phlogiston', together with compounds like 'dephlogisticated air' derived from it, is one of the ways by which the original text communicated the beliefs of its author. Substituting unrelated or differently related expressions for those related, sometimes identical terms of the original must at least suppress those beliefs leaving the text that results incoherent ... To be coherent a text that deploys the phlogiston theory must represent the stuff given off in combustion as a chemical principle, the same one that renders the air unfit to breathe and that also, when abstracted from an appropriate material, leaves an acid residue. (1983, pp. 675-6)

In her discussion of the debate between Kitcher and Kuhn, Hesse adds the comment that:

> We have not only to say that phlogiston sometimes referred to hydrogen and sometimes to absorption of oxygen, but we have to convey the whole ontology of phlogiston in order to make plausible why it was taken to be a single natural kind. (1983, p. 707)

My argument above that contextual translation obliterates the implication of a kind parallels Kuhn's objection. However, Kuhn goes on to make a specific suggestion about what translation must preserve. This suggestion is problematic.

Kuhn sketches his view of what he calls "the invariants of translation" (1983, pp. 681-3). "Different languages", he says, "impose different structures on the world" (p. 682). Their lexicons have a structure which "mirrors aspects of the structure of the world which the lexicon can be used to describe". Such "lexical structures" may be "homologous"; homologous structures "mirror [ ] the same world" (p. 683). "What such homologous structures preserve ... is the taxonomic categories of the world and the similarity/difference relationships between them." Homology of "taxonomic structure" is the key to "the invariants of translation":

> speakers of mutually translatable languages need not share terms ... the referring expressions of one language must be matchable to coreferential expressions in the other, and the lexical structures ... must be the same ... from one language to the other. Taxonomy must ... be preserved to provide both shared categories and shared relationships between them. (1983, p. 683)

So far this is unobjectionable. Kuhn's demand for the preservation of "taxonomic structure" agrees with my argument that translational expressions must represent the same kind. For in order to translate a term for a kind from one theory into another it seems necessary that the theories possess the same classificatory or taxonomic system.

Problems arise with Kuhn's discussion of the relation between taxonomic structure and reference determination. Kuhn implies that it is not how reference is fixed that must be preserved in translation, but taxonomic structure alone. That is, a pair of languages may be intertranslatable by homology of taxonomy even if the reference of their terms is not determined in the same way.

Kuhn bases this view on the point that "different people use different criteria in identifying the referents of shared terms" (p. 681). He says "'criteria' is to be understood in a very broad sense, one that embraces whatever techniques, not all of them necessarily conscious, people do use in pinning words to the world" (pp. 685-6, fn. 13). He explains the relation between terms and criteria within individual idiolects by saying that:

> for each individual [speaker] a referring term is a node in a lexical network from which radiate labels for the criteria that he or she uses in identifying the referents of the nodal terms. (1983, p. 682)

To connect the idea with taxonomy, he says "homologous [lexical] structures ... may be fashioned using different sets of criterial linkages" (p. 683). For speakers who share a language, the "criteria need not be the same", but they must "share ... homology of lexical structure", "their taxonomic structures must match".

From speakers of the same language not sharing criteria to the intertranslatability of languages which do not share criteria is no great step. To share a common language, individual speakers need only share taxonomy, not criteria. Similarly, intertranslatable languages need not share criteria, only taxonomy.

But it is a step that can only be taken if the relation between languages is the same as that between individual speaker's idiolects. The trouble is that while it can be explained why speakers of the same language need not individually possess all criteria, it is not at all clear that sameness of taxonomy across languages is sufficient for translation in the complete absence of criteria in common. That individual speakers of a language do not need to share criteria follows from Putnam's "socio-linguistic hypothesis" of the "division of linguistic labour" (1975b, pp. 227-9). Putnam's idea is that a speaker need not know how a given term's reference is determined, since reference can be borrowed from other speakers who do possess the means to fix or identify its referents. Kuhn's idea is a minor departure from Putnam's: different speakers may use a term to refer to the same kind even though they do not themselves possess all the same ways in which its reference is fixed.

Kuhn does not explicitly state that criteria need not be shared by intertranslatable languages. But, as we have seen, he is perfectly explicit that intertranslatable languages must share "lexical structures", and that "taxonomy must ... be preserved". And he plainly states that "homologous structures ... may be fashioned using different sets of criterial linkages". If homology of structure suffices for translation and homologous structures need not share criteria, then he is committed to the possibility of intertranslatable languages which do not share any criteria.

In order to make the following argument more precise it is necessary to single out a specific function of Kuhn's criteria of reference. Kuhn takes a criterion to be what speakers use to identify or pick out referents. No doubt some of the ways speakers identify referents are mere incidental beliefs and need be shared neither between speakers nor between languages. However, some ways of picking out referents must also have the role of fixing reference.

On the assumption that Kuhn's "criteria" include determinations of reference, it follows that the terms of intertranslatable languages need not have their reference determined in the same way. And, in particular, a term from one language could be translated into another even if the reference of its translation were not determined in the same way in the other language.

This consequence of Kuhn's position can be shown to rest on a fallacy. The consequence derives from two assumptions made by Kuhn. We have seen that it is necessary for translational expressions to refer to the same kinds. But, since Kuhn takes sameness of taxonomy to suffice for translation, he further assumes that reference to the same kind is sufficient for translation. And since he takes sameness of taxonomy not to require sameness of reference determination, Kuhn also assumes that it is possible to refer to the same kind without determining reference in the same way. The latter assumption may be granted, since reference can be secured to the same kind by different means of determining reference. Yet it is a fallacy to assume that reference to the same kind is sufficient for translation. For, as argued in section 3.3, different means of determining reference give rise to difference of semantic content. Thus sameness of taxonomy fails to insure that expressions which refer to the same kinds are semantically equivalent. It fails, therefore, to yield a sufficient condition of translation.

Notes

1. As with, e.g., Devitt (1984, p. 136).
2. Many authors, following Carnap (1956) call possible reference 'intension'. That use is avoided here because 'intension' can also mean sense.
3. The example is Putnam's (1981, p. 27). He cites it as an objection to taking Carnap's notion of intension — i.e. extension in all possible worlds — as an analysis of meaning.
4. Cf. Feyerabend (1981e, pp. 99-100), (1981h, pp. 153-4), (1965, pp. 168-70).
5. As an aside, note Claggett's remark that "there is no possibility of an inertial concept in a system that demands continuing force wherever there is motion" (1959, p. 456, fn. 8). This sits well with Feyerabend's idea of the inability to form a concept on the basis of an opposing system of concepts.
6. Feyerabend (1981d, p. 65) discusses such measurement and refers to Claggett, whose comment is worth quoting: "...the impetus ... seems close to being the effectiveness which the original force has on a particular body, an effectiveness measureable in terms of the velocity immediately supplied to

the body and the quantity of matter in the body ... the terms of its measure as presented by Buridan make an analogue with momentum, i.e., this quality which is motive force for Buridan turns out to be described in dimensions analogous to those of Newton's momentum" (1959, p. 523).

7. See the debate between Kuhn (1983) and Kitcher (1983). In addition to these two articles, the discussion derives from Conant (1964), Hankins (1985) and Le Grand (1987).
8. Kitcher (1978, pp. 529-35); cf. (1982, pp. 339-40).
9. Mercury, incidentally, is the only metal whose calx converts back into the metallic state by heat alone, without charcoal. That is why oxygen, rather than carbon dioxide, is obtained in the reaction.
10. Kitcher never clearly states whether mere co-reference is all that is required to specify the reference of a term-token using some other. But he takes 'oxygen' as the translation of some tokens of 'dephlogisticated air', which suggests that it is. Yet if co-reference is all that is required, then his contextual translations are based on the fallacy that co-reference suffices for translation.
11. Of course, if a "blank" as Kuhn calls it (1983, p. 674) appears in the place of the token of 'dephlogisticated air' to be translated, then contextual translation is no translation at all.
12. Strictly speaking, an ambiguous term-type is not a semantical term-type but an inscription type. The ambiguity is because a single inscription type symbolizes distinct semantic types.
13. Similar remarks apply to an argument of Levin's (1979, pp. 410-1). He adopts Field's (1973) idea that Newtonian 'mass' partially denotes two sorts of mass, arguing that it is ambiguous and should divide in translation. That takes diffuse reference as a criterion of ambiguity. But clearly a term may partially denote two different referents without being ambiguous.

4 In defence of untranslatability

4.1 Introduction

This chapter addresses criticisms of the concept of untranslatability which Davidson (1984) and Putnam (1981) have raised against the incommensurability thesis.

The main themes of the criticism are present in the following extract from Putnam (1981):

> The incommensurability thesis is the thesis that terms used in another culture, say, the term 'temperature' as used by a seventeenth-century scientist, cannot be equated in meaning or reference with any terms or expressions we possess ... [I]f this thesis were really true then we could not translate other languages — or even past stages of our own language — at all. And if we cannot interpret organisms' noises at all, then we have no grounds for regarding them as thinkers, speakers, or even persons. In short, if Feyerabend (and Kuhn at his most incommensurable) were right, then members of other cultures, including seventeenth-century scientists, would be conceptualizable by us only as animals producing responses to stimuli (including noises that curiously resemble English or Italian). To tell us that Galileo had 'incommensurable' notions and then to go on to describe them at length is totally incoherent. (1981, pp. 114-5)

The central objection is the claim that it is incoherent to talk about what is untranslatable. Three lines of argument may be distinguished with regard to this alleged incoherence.

One is a direct argument against untranslatability. Put simply, it is incoherent to express the content of a purportedly untranslatable language within the language into which it is said to be untranslatable. This argument will be dealt with in section 4.2.

The other two arguments are indirect arguments which proceed by way of the assumption that translation is necessary for understanding. One is that it is incoherent to profess to understand ideas expressed in an untranslatable language because they are incomprehensible. The other is that it is incoherent to think of the speaker of an untranslatable language as having a language at all. These two arguments are the topic of 4.3.

In addition to incoherence arguments, Davidson (1984) argues that languagehood is inextricable from translation. He claims that a "dogma of a dualism of scheme and reality" which fallaciously separates language from translation underlies the incommensurability thesis. His attack on the dualism will be considered in sections 4.4 to 4.6.

4.2 The direct incoherence argument

Putnam defines 'incommensurability' by saying that "terms used in another culture ... cannot be equated in meaning or reference with any terms or expressions we possess". Given this definition of incommensurability, the direct incoherence argument is embodied in the last sentence of the quote: "To tell us that Galileo had 'incommensurable' notions and then to go on to describe them at length is totally incoherent". For if Galileo's ideas really are untranslatable into our language, then they cannot be expressed using our language, and it contradicts the claim of untranslatability to do so.

Davidson puts the point in the form of a paradox (1984, pp. 183-4). "We are encouraged", he says, to "imagine we understand massive conceptual change" by the use of examples, but "the changes and the contrasts can be explained and described using the equipment of a single language". "Kuhn", he adds, "is brilliant at saying what things were like before the revolution using — what else? — our post-revolutionary idiom" (1984, p. 184). The paradox is that the meaning expressed by the terms of an untranslatable language should be expressed using the very language into which translation allegedly fails.

Thus Davidson and Putnam give only a sketch of the direct argument. The argument is a meta-argument because it specifically applies to arguments for untranslatability. We may specify it more fully as follows. Suppose it is argued in language L that L* is untranslatable into L. Suppose as well that the argument in L employs examples from L* in the sense that it expresses the meaning of terms taken from L*. It follows from the latter that L* is translatable into L, for that is what expressing the meaning of terms from L* in L amounts to. But then the argument itself translates from L* into L in the course of arguing that L* is not translatable into L. If the argument is correct, then it is possible to translate from L* into L. But in that case the conclusion of the argument is false. If the conclusion is correct, then it is impossible to translate from L* into L. But in that case the argument is incorrect, for it assumes translation between L* and L. An argument in which the truth of the premises is incompatible with the truth of the conclusion is incoherent.

The meta-argument is sound but its scope is limited. It is neither a general objection to the idea of untranslatability as such, nor is it a general criticism applicable to all arguments for untranslatability. Rather, it is a restricted argument whose sole function is to show that one particular form of argument for untranslatability is self-refuting.

In the first place, the immediate target of the objection is not untranslatability as such. The objection does not show that the actual relation of untranslatability is an incoherent relation. Nothing follows about untranslatability itself from the fact that it is incoherent to translate what is untranslatable. Such incoherence applies only to the possibility of translating the untranslatable, and entails nothing about the possible instantiation of the relation of untranslatability by actual languages. Far from being a criticism of the untranslatability relation itself, the objection is specifically directed against arguments for untranslatability.

In the second place, it is not even a general criticism of all arguments for untranslatability. It is a criticism of only those arguments in which the untranslatability argued for attaches to the language of argument. It is indexed exclusively to arguments in which the language of argument is the language into which translation fails. Such an objection can have no point of entry where the language of argument and the untranslatable languages are distinct. As long as the semantic analysis of a pair of object-languages is couched in a metalanguage distinct from either object-language, their semantic content can be compared without translating from one into the other. Thus, it can be shown in a metalanguage that a pair of object-languages is not intertranslatable without expressing untranslatable content in either object-language.

In the third place, the criticism applies only to arguments which employ examples.[1] Arguments for untranslatability not dependent on examples are immune to such criticism. For instance, an argument deriving from general considerations in the theory of meaning need not make use of examples. One such argument is that meaning in general, and the meaning of scientific terms in particular, depends on context, and hence sufficient difference in theoretical context makes scientific terminology untranslatable. It is irrelevant whether such an argument is defensible independently. The point is that it is possible to argue for untranslatability without using examples.

These three points show that the direct incoherence argument is not fully general. The argument does not, therefore, show the notion of untranslatability to be self-contradictory. If it did show that, any claim of untranslatability could be dismissed out of hand. But since the argument is not general, it cannot be brought to bear on any particular claim of untranslatability unless it is specifically shown to apply to it.

Of course, the reason Davidson and Putnam employ the incoherence objection in the first place is that they assume incommensurability falls within the ambit of the argument. They assume that the language into which an untranslatable theory fails to be translatable is the language in which the argument for incommensurability is couched. Instead of translation failure between delimited theoretical terminologies, they identify the language into which translation fails with language as a whole. In part, they assume this because they take the issue to be translation of the vocabulary of an old abandoned theory into contemporary natural language.

This interpretation is explicit in Putnam's definition of "the incommensurability thesis [as] the thesis that terms used in another culture ... cannot be equated in meaning or reference with any terms or expressions we possess". And it is evident in the inference he draws: "if this thesis were really true then we could not translate other languages ... at all". When Davidson notes Kuhn's paradoxical use of "our post-revolutionary idiom" to discuss pre-revolutionary science he assumes that the modern language into which out-of-date theory fails to translate is contemporary English. Davidson also seems to take the language into which translation fails to be a total language because he discusses incommensurability in the course of his analysis of complete translation failure (1984, pp. 190-1). As Davidson and Putnam interpret incommensurability, the language of argument and the language into which translation fails are one and the same.

It remains only to note that Kuhn and Feyerabend make extensive use of examples, both in the course of their arguments and in independent discussion. Their handling of incommensurability is thus open to the charge of incoherence. For if the language into which translation fails is the very language in which untranslatability is argued for, then their use of examples is indeed incoherent.

Thus the Davidson-Putnam argument depends on two main premises. First, it assumes that the language in which it is argued that a theoretical language is untranslatable is the very language into which translation fails. Second, it assumes Kuhn and Feyerabend give the meaning of the terms which they claim are untranslatable.

This poses a dilemma. To defend the idea of untranslatability between theories against the incoherence objection, at least one of the premises must be denied. It must be denied either that the language into which translation fails is the language in which the argument is couched, or that Kuhn and Feyerabend give the meaning of their examples of untranslatable terms.

Consider the latter option. This amounts to saying that when Kuhn and Feyerabend discuss such examples as 'impetus' and 'dephlogisticated air' they fail to express their meaning. Instead they give partial translations or mere specifications of reference for such expressions.

But this option is unacceptable. To downplay Kuhn and Feyerabend's treatment of examples in this way would be a serious misrepresention. Their exposition of the meaning of expressions is what shows the expressions to be untranslatable in the first place. It must be denied instead that the language into which translation fails is the language of argument. To do this, it must be denied that incommensurability entails untranslatability into a total language.

The latter denial is in full accord with incommensurability. For the incommensurability of scientific theories has nothing to do with relations between total languages. Although it is a relation between the languages of theories, the language specific to a theory is only a part of a language, and is not to be identified with language as a whole.

The incommensurability of theories is due to semantical differences in the terminology or vocabulary theories employ. More exactly, the terminology employed within one theory cannot be translated into the terminology of a theory with which it is incommensurable. Thus, instead of untranslatability into a total language, it is rather a case of translation failure between sub-languages or local idioms within language as a whole.

That the untranslatability is thus limited and localized is evident from Kuhn's and Feyerabend's discussions of the theories they take to be incommensurable. In their discussions no mention is made of total languages, since they are exclusively concerned with semantic analysis of the vocabulary the theories employ.[2]

Kuhn makes the point explicitly in connection with the Davidson-Putnam argument (1983, pp. 669-71). He advocates "local" incommensurability, which is untranslatability between sub-sets of the terms used by a pair of theories. This involves localized semantic difference within the context of generally shared everyday and scientific language.

> Most of the terms common to ... two [incommensurable] theories function the same way in both; their meanings ... are preserved; their translation is simply homophonic. Only for a small subgroup of (usually interdefined) terms and for sentences containing them do problems of translatability arise. (1983, pp. 670-1)

Thus Kuhn's view is not even that there is full translation failure between the special languages of incommensurable theories. Rather, each theory has a central complex of terms which is interdefined as a whole, and which are not translatable in whole or in part into a theory for which that particular "local holism" does not obtain.

The situation is the same with Feyerabend. It is less straightforward since he applies the concept of incommensurability more generally, including within its ambit languages and world-views as well as theories.[3] This can confuse the issue: thus Putnam in the quotation which opens this chapter runs together the idea that theories are incommensurable with the idea that culturally embedded languages are.

But added generality does not affect the issue. Widened application does not imply that the vocabulary of incommensurable theories is untranslatable into a broader natural language. What follows is instead that the set of relata which enter the incommensurability relation includes more than just scientific theories. This does not affect the nature of the relation between pairs of incommensurable theories. As far as their incommensurability is concerned, it is a matter of indifference whether languages or world-views may enter similar relations among themselves or even that a theory might be incommensurable with a world-view or language.

In discussing incommensurability as a relation between scientific theories, Feyerabend characterizes it exclusively in terms of relations between theories. As distinct from Kuhn's notion of clusters of interdefined terms, Feyerabend holds that the basic principles of a theory preclude the formation of the concepts of a theory with which it is incommensurable.[4] The resultant untranslatability affects more than a cluster of terms, but is still a relation between the languages of theories, rather than total languages.

Theoretical sub-languages constitute fragments of an inclusive natural language. The picture of language which thus emerges is of natural language as a conglomerate of terminologies or local idioms with special areas of application. Untranslatability between theoretical languages therefore constitutes a relation between sub-languages within a total language.

Rather than untranslatability into a total language as assumed by Davidson and Putnam, what is at issue is localized translation failure between sub-languages contained in a total language. Thus the language into which the vocabulary of a theory fails to be translatable may very well be distinct from the language of argument. For the argument that a pair of sub-languages is not intertranslatable can be couched in a portion of language distinct from the language into which translation fails. Theoretical sub-languages may themselves be the topic of a discussion carried out within some other fragment of the language.

Let us consider two languages TL and TL* associated with two theories. We may suppose that TL and TL* are sub-languages of a broader natural language L. It is possible to use L as a metalanguage to speak about the semantic relations between TL and TL*. In particular, it may be argued in L that a term t* of

TL* cannot be translated into TL. Such an argument need not be formulated in TL, for it can be formulated in L. Using L as metalanguage, t* can be referred to and shown to be indefinable in TL without being expressed in TL. Nor is there any need in the course of the argument to express the content of t* in TL. For t* may be defined in L used as a metalanguage for TL without formulating the content of t* in TL.

To make the point concrete, consider the discussion of 'impetus' in 3.5. Following Feyerabend, I noted that the concept of impetus is indefinable within Newtonian mechanics. Impetus is supposed to be a cause which acts constantly upon all projectiles, whereas the Newtonian theory denies that inertial motion is subject to a sustaining cause.

The discussion concerns semantical relations between the vocabularies of impetus theory and Newtonian mechanics. The discussion is couched in a fragment of English employed as a metalanguage. It mentions 'impetus' and gives its meaning as defined in the impetus theory. Then it is pointed out that such a definition cannot be formulated on the basis of the principles of Newtonian mechanics. This is a point about the semantical limitations of the language of Newtonian mechanics, and the point is made in the metalanguage. The meaning of the term 'impetus' is not expressed within the language of the Newtonian theory at any place in the argument. For the language of argument and the language into which 'impetus' fails to be translatable are distinct.

In sum, the direct incoherence argument does not apply to the incommensurability thesis. Given that the untranslatability in question is a restricted relation between theoretical sub-languages, and given that such sub-languages may be discussed within a metalanguage, no incoherence attaches to the untranslatability argument. For the language in which the untranslatability is argued for is not the language into which translation fails. We may conclude that Davidson and Putnam fail to directly refute the incommensurability thesis.

4.3 Translation and interpretation

In addition to the direct argument, the Putnam passage quoted in 4.1 suggests arguments which do not proceed strictly in terms of translation. The relevant excerpts are the following.

> ...if [the incommensurability] thesis were really true then we could not translate other languages ... at all. And if we cannot interpret organisms' noises at all, then we have no grounds for regarding them as thinkers, speakers, or even persons ... seventeenth-century scientists [ ] would be conceptualizable by us only as animals producing responses to stimuli... To tell us that Galileo had 'incommensurable' notions and then to go on to describe them at length is totally incoherent. (1981, pp. 114-5)

This passage suggests two main lines of argument: that the meaning of untranslatable expressions cannot be known; and that an organism whose language is untranslatable cannot be known to be a speaker.

Both arguments depend on the assumption that a language which cannot be translated cannot be interpreted. This assumption is implicit in Putnam's inference from "we could not translate ... at all" to "we cannot interpret organisms' noises at all". That discussions of incommensurability are incoherent purportedly follows from failure of interpretation.

It is not immediately clear what Putnam means by 'interpret'. To say that translation failure entails inability to "interpret organisms' noises" suggests that the meaning of untranslatable expressions cannot be understood. To interpret an expression is presumably to understand it, i.e. to construe its meaning in a particular way. But how is the concept of interpretation related to that of translation?

Putnam may assume that interpretation of a speaker who shares one's own language constitutes homophonic translation from the speaker's idiolect into one's own idiolect. And if domestic interpretation is conceived in this way — in effect, as a form of translation — it is natural to conceive interpretation of foreign language as necessarily a translational process as well. On such a conception, interpretation of foreign language expressions would consist in understanding translationally equivalent expressions within a home language. In this translational sense of interpretation, interpretation of a foreign language is a twofold process. It consists in translating foreign expressions from a foreign language into a home language and understanding their home language equivalents. Thus conceived, failure to translate immediately entails failure of interpretation.

This sense of interpretation can be given a weak or a strong construal, depending on whether the translational component involves exact or loose translation. If the translational component must be exact, interpretation of foreign language expressions would consist in understanding their semantic equivalents in a home language. If the standard of translation is relaxed, interpretation might consist in understanding a loose rendering of the meaning of foreign expressions.

If 'interpret' were given the first, stronger, reading, Putnam's argument would fail. It does not follow from failure of exact translation that the content of speakers' utterances cannot be understood. For that to follow, it would at least have to be the case that exact translation is a necessary condition for understanding such content. But there is no reason to assume that failure of exact translation entails that a language cannot be understood at all. The language could be learned directly, without the aid of a translation. Moreover, failure of exact translation need not preclude the production of a gloss or loose translation. Over and above mere failure of exact translation, what is needed is for comprehension of meaning to fail altogether.

To take interpretation in the second way as loose translation is to implausibly exaggerate incommensurability. Untranslatability of theories in the sense relevant to incommensurability does not entail the total absence of common semantic features: expressions of untranslatable languages may share some aspects of reference and even meaning.[5] Though 'dephlogisticated air' cannot be translated into the oxygen theory, some of its tokens co-refer with 'oxygen'. A loose rendering of phlogistic concepts in the terminology of the oxygen theory is not ruled out by untranslatability. If 'interpretation' is taken in a loose sense, Putnam's denial of interpretation is stronger than licensed by incommensurability.

In the context of the argument, therefore, interpretation cannot itself be taken to consist in translation. Interpretation must be distinct from translation. Though interpretation may in some cases depend on translation, it cannot have translation necessarily built in to it as a constitutive component.

To interpret an expression must simply be to understand what it means. To understand an expression is not to translate it, nor is understanding retricted to what is expressed in a home language. Rather, to understand consists simply in knowing the meaning of an expression, whatever language it belongs to.

Putnam's inference from failure to translate to failure to interpret does not require that interpretation consist in translation. It suffices for that inference if it is assumed that translation is a necessary prerequisite of understanding or interpreting a foreign expression. For translation to be necessary for interpretation is distinct from its being a component of the process of interpretation.

The assumption that translation is necessary for interpretation is a restrictive assumption about understanding, which differs from the view that interpretation is itself a form of translation. It is an assumption that understanding is limited to expressions couched in one's home language. If the assumption were true, we would be unable to come to know the meaning of an expression not translatable into our language. The point is not that such concepts would be beyond our psychological capacities of comprehension, but that epistemic access to them would be blocked by failure to express their content in our language.

The first of Putnam's two arguments depends on the thesis that we cannot know what is expressed in a language which we cannot translate. The argument is as follows. Assume that translation is necessary for interpretation, so that it cannot be known what expressions in an untranslatable language mean. Now notice that advocates of the incommensurability thesis do profess to understand the meaning of expressions in languages which they claim to be untranslatable. Thus they say both that the expressions cannot be translated and that they know what the expressions mean. But this is incoherent: for if the expressions cannot be translated, their meanings cannot be known; and if the meanings are known, then the expressions can be translated.

The second argument is that it makes no sense to attribute ideas to an organism unless there is evidence of language possession. Assume as before that translation is necessary for interpretation. If a speaker's utterances cannot be translated, then it cannot be known what the utterances mean. If no meaning can be attributed to the utterances, then there is no evidence the speaker has a language. Advocates of incommensurability both describe speakers as having untranslatable languages and attribute meanings to the speakers of such languages. But that is incoherent: for if utterances cannot be translated there is no evidence the speaker has a language; and if meaning is attributed to the utterances, that presupposes the speaker does have a language.

The first argument derives immediately from the assumption that translation is necessary for interpretation. If translation is necessary for interpretation, then it is self-contradictory to profess to understand an untranslatable expression. The second argument depends on that assumption and reaches a similar conclusion. But it does so, not simply by way of the denial that untranslatable content can be understood, but by denying there is any reason to take a speaker whose utterances are uninterpretable to have a language. The point of the second argument is that it is incoherent both to take an organism to be devoid of language and to ascribe meaningful utterances to it.

The assumption that translation is necessary for understanding another language is implausible for a number of reasons. First, there is no reason to assume that understanding a foreign expression consists in understanding its translational equivalent within the home language. Bilingual speakers do not translate "in their heads" while conversing in a foreign language. Hence there is no reason to suppose that a bilingual could not understand a foreign expression not translatable into his home language. Second, if understanding another language really did require translation into a prior language, it would be impossible to learn a first language. Presumably, a child does not come into the world already equipped with a language, but must learn its first language directly. Third, translation is unnecessary for learning a second language. It is possible to learn a new language as a child learns its first one by the method of direct immersion. It is perhaps unavoidable to assign meanings of home language terms to apparent equivalents in the new language. But that is no reason to suppose there could not be expressions understood in the new language, which are inexpressible in the home language. Fourth, if translation were necessary for understanding a foreign language, then it would be completely mysterious how translation could ever take place. For one can only translate a language if one understands the language. If translation were necessary for understanding, then it is quite unclear how the language could ever be understood in the first place.[6]

The preceding points are patterned on the responses of Kuhn (1983) and Feyerabend (1987). Both claim that the language of a theory incommensurable with one's own can be understood. Feyerabend rebuts Putnam by pointing out that "we can learn a language or a culture from scratch, as a child learns them, without detour through our native tongue" (1987, p. 76). Kuhn distinguishes between translation of a language and interpretation of an initially unintelligible language. He characterizes interpretation as follows:

> Unlike the translator, the interpreter may initially command only a single language. At the start, the text on which he or she works consists in whole or in part of unintelligible noises or inscriptions ... If the interpreter succeeds, what he or she has in the first instance done is learn a new language ... whether that language can be translated into the one with which the interpreter began is an open question. Acquiring a new language is not the same as translating from it into one's own. Success with the first does not imply success with the second. (1983, pp. 672-3)

This distinction enables Kuhn to rebut Putnam as Feyerabend does. For an untranslatable theory may be interpreted, so there is nothing incoherent about claiming to understand the meaning of untranslatable expressions.

Thus, the distinction between understanding or interpreting a language and translating it into one's own permits a direct rebuttal of the first argument. This rebuttal may be further elaborated by noting that the untranslatable language to be interpreted in the case of an incommensurable theory is not a total language. Incommensurability does not entail radical translation. Rather, what is at issue is untranslatability within a single language, and what must be interpreted is an unknown area of that language.

Not to make this point would allow the required interpretation appear to pose too great a task. I noted in 4.2 how Putnam exaggerates the scope of the languages affected by the untranslatability of theories. He takes it to be a relation between total languages, which makes discussion of untranslatable expressions paradoxical. A similar exaggeration may increase the plausibility of the assumption that translation is necessary for understanding and make our denial of that assumption appear facile. For as a practical matter, discourse in a language of which one is ignorant cannot be understood unless one has a translation manual. Short of learning such a language directly, inability to translate prevents comprehension. So, if the interlinguistic situation between incommensurable theories is one of radical translation, direct language acquisition may be impracticable. Untranslatability of theory would entail pragmatic incomprehensibility, and the possibility of direct interpretation would constitute an irrelevant abstraction.

But it is misconceived to take incommensurability to entail radical translation. For, as noted in 4.2, the terminologies of incommensurable theories constitute sub-languages embedded in a common natural language. Rival theorists who espouse incommensurable theories may therefore share a background language. Untranslatability between their theories obtains within the context of ongoing communication in a shared language. Problems of comprehension of untranslatable theoretical terminology are therefore localized within a common language which provides the basis for discussion of the areas of difficulty. Thus the linguistic situation is more akin to learning a new jargon than confrontation with a whole new language.

Interpretation of unknown vocabulary untranslatable into one's own theory is therefore not the radical project of learning a completely unknown language without the benefit of any common language. Indeed, there is a broad background of shared language which may be used to assist in interpretation. The background language may even be employed as a metalanguage to discuss unknown theoretical terminology. Without translating between incommensurable theoretical languages, explanation of the meaning of their terminology may take place in the metalanguage.

To illustrate, consider 'impetus' again. The term 'impetus' may be discussed in English. The discussion is metalinguistic because it concerns semantic features of 'impetus'. Semantic features which prevent 'impetus' from being defined within Newtonian mechanics are explicable in such a discussion. For example, the causal role description which fixes the reference of impetus as an indwelling force which sustains projectile motion can be expressed in English even though it is not formulable on the basis of Newton's laws. Thus 'impetus' can be explained in English even though it cannot be translated from the idiom of the impetus theory into the Newtonian idiom.

Let us turn to Putnam's second argument, which is that untranslatability prevents language attribution. The preceding discussion disposes of the initial premise that an untranslatable language cannot be understood. But we may also question its main inference that "if we cannot interpret organisms' noises at all, then we have no grounds for regarding them as thinkers, speakers, or even persons".

The argument would succeed if it were true that an untranslatable language cannot be recognized as a language. It is undeniable that it is incoherent to deny possession of a language to an organism while saying of that organism that it possesses concepts which it expresses in language. What should be questioned instead is whether a speaker whose language cannot be translated cannot be known to possess a language. If a language can be recognized without translating, then it would not follow from inability to translate that it is incoherent to attribute the ability to express concepts.

Putnam apparently assumes that if the meaning of sounds or inscriptions cannot be interpreted then there is no reason to take the organism which produces them to have a language. Davidson takes a similar view when he asks us to reflect "on the close relations between language and the attribution of attitudes":

> On the one hand, it is clear that speech requires a multitude of finely discriminated intentions and beliefs... On the other hand, it seems unlikely that we can intelligibly attribute attitudes as complex as these to a speaker unless we can translate his words into ours. There can be no doubt that the relation between being able to translate someone's language and being able to describe his attitudes is very close. (1984, p. 186)

With both Putnam and Davidson, the suggestion appears to be that knowledge of meaning or propositional attitude is required to justify language attribution.

This suggestion is surely mistaken. Why should knowledge of semantic content be necessary for language recognition? Surely, formal and contextual features count for something. Codes may be recognized as codes without being broken. Similarly, fragments of dead languages (e.g. linear B, hieroglyphics) may be recognized as such prior to translation. Tourists typically recognize native speech as the local tongue even if they do not understand it. And why must psychological content be determined to identify behaviour as linguistic? In many social and physical settings the observed behaviour of humans is identifiable as linguistic without access to attitude or meaning. In any case, mental state need not be entirely inscrutable in the absence of knowledge of a language. The rough character of attitude or meaning can be known from observation of non-linguistic aspects of behaviour. In sum, language has structure and linguistic behaviour is enmeshed with practical activity and social relationships in such a way that non-semantic features of language use permit identification of language.[7]

Even if there were no way to determine the presence of language without access to attitude or meaning, it would still not follow that an untranslatable language could not be identified as a language. The presumed necessity of content ascription by means of translation presupposes the necessity of translation for interpretation. To say that a speaker's meanings or attitudes can only be known if the speaker's language can be translated is to assume the only way to understand is via translation. But the possibility of direct understanding or acquisition of a language means that meaning and belief are interpretable without translation. A bilingual may determine psychological and semantical content for speakers of an untranslatable language without translating back into a home language.

In any case, the problem of recognizing language is largely irrelevant to incommensurability. A rival theorist is not an organism whose possession of language is in question. Intertheoretic untranslatability is a relation restricted to terminologies within an inclusive language. So scientists with untranslatable theories may share a natural language. The problem of recognizing a speaker as having a language does not even arise. Whether a rival theorist actually possesses a language is an issue resolved prior to discussion of theory. Nor is it as if the discovery of semantic variance between theories throws into question the status of a scientist as a speaker of language. For shared use of a background language is a precondition of narrowing a linguistic difference down to difference of theory.

This completes criticism of Putnam's two arguments. I will now briefly consider a related argument which derives from Davidson's discussion of interpretative charity (1984, pp. 195-7).[8] For Davidson, interpretation of a speaker requires charity and charity implies translation, so interpretation entails translation. Success in interpretation is therefore inconsistent with translation failure. Thus the incommensurability thesis may seem incoherent because it presupposes successful interpretation of scientists whose theories are not translatable into ours.

Davidson applies the principle of charity to the problem of radical interpretation. The problem is how to interpret meaning without independent access to belief: "a man's speech cannot be interpreted except by someone who knows a good deal about what the speaker believes ... and ... fine distinctions between beliefs are impossible without understood speech" (1984, p. 195). He assumes that "the basic evidence for a theory of radical interpretation ... [is] the attitude of accepting as true, directed to sentences". But such evidence does not determine meaning: "if we merely know that someone holds a certain sentence to be true, we know neither what he means by the sentence nor what belief his holding it true represents" (p. 196). Charity is invoked to extract meaning from the thin evidence of sentences held true.

To determine meaning, assumptions must be made about belief: "if all we know is what sentences a speaker holds true, and we cannot assume that his language is our own, then we cannot take even a first step towards interpretation without knowing or assuming a great deal about the speaker's beliefs" (p. 196). Belief attribution should be governed by charity. The rough idea is for the agent to come out on the whole as a believer of truths:

> We get a first approximation to a finished theory by assigning to sentences of a speaker conditions of truth that actually obtain (in our opinion) just when the speaker holds those sentences true. The guiding policy is to do this as far as possible, subject to considerations of simplicity, hunches about the effects of social conditioning, and of course our common-sense, or scientific, knowledge of explicable error. (1984, p. 196)

The principle of charity is justified because the agreement it provides is a precondition of interpretation.

> Since charity is not an option, but a condition of having a workable theory, it is meaningless to suggest that we might fall into massive error by endorsing it. Until we have successfully established a systematic correlation of sentences held true with sentences held true, there are no mistakes to make. Charity is forced on us; whether we like it or not, if we want to understand others, we must count them right in most matters. (1984, p. 197)

The connections Davidson draws between interpretation, translation and charity seem to license the following inferences. Since charity involves taking sentences of our language which we hold true as the content of alien utterances, charity implies translation. Since charity is necessary for interpretation, successful interpretation entails translation. Therefore, interpretation of an agent is inconsistent with translation failure. Thus to interpret a scientist as having a theory untranslatable into one's own is incoherent.

In effect, Davidson's use of the principle of charity combines Putnam's two arguments. In accordance with the first argument, charity makes translation necessary for interpretion. In accordance with the second, it makes translation necessary for interpreting an agent as a speaker. Two objections may be raised against this use of the principle of charity.

In the first place, the link between charity and translation must be severed. Just as interpretation does not require translation, interpretative charity does not require translation. Davidson assumes that charitable interpretation of an agent assigns truth conditions in a home language to sentences of an alien language. But while such charity might be generally advisable, it is not necessary. Charity may be applied directly within the alien language. Charity may be incorporated into the direct method of language acquisition. In learning a language directly without translating the interpreter can, and perhaps should, assign maximum plausible truth conditions as well as reasonable belief. Interpretation seeks coherence and assigns plausible truth-values whether or not it results in a translation.

In the second place, charity is unsuitable for theoretical discourse. The principle of charity can be refined in various ways to allow for varying degrees of error. But the general principle of assigning maximal truth is unacceptable as a principle of interpretation when applied to theoretical languages.[9] Maximal assignment of truth to the statements of a scientific theory overlooks the possibility of large-scale error. But the history of science shows that theories have frequently been profoundly mistaken. Moreover, there are compelling epistemological reasons to take a fallibilist stance towards all theories, past and present.[10] Surely, in the interpretation of scientific language no assumption about the truth of theoretical assertions should be made.

Now, against this second objection, it might be argued that attribution of massive error makes behaviour unintelligible. That is, to deny of an agent that any of its beliefs are true is to make it inexplicable how it manages to engage in successful action. But to say that a theory is totally or mostly false is not to say that the entirety of an agent's beliefs are false. To deny that the theoretical claims of the phlogiston theory are true does not make it impossible to explain how Priestley engages in practical action. Moreover, a false theory can have true consequences and be put to practical use. And a theory which is strictly false but nearly or approximately true may serve well as a guide for action.

Perhaps Davidson intends to exempt theory from maximal assignment of truth. For he does say we should assign truth "subject to considerations of ... common sense, or scientific, knowledge of explicable error". But he fails to elaborate the point. In any case, the general policy of overall interpretative charity towards speakers should not be enjoined upon the interpreter of theoretical discourse. For the purpose of interpreting theoretical discourse, we are not therefore obliged by the principle of charity to impose translational equivalences upon scientific theories. So the forcing move from charity to intertheoretic translatability may be rejected. The possibility of interpretation does not rule out translation failure between theories.

4.4 The scheme-content dualism

There is another side to Putnam's claim that we have no reason to take uninterpretable organisms as "thinkers, speakers, or even persons". Namely, putative linguistic activity whose meaning is uninterpretable is indistinguishable from non-linguistic behaviour. The point is clearer with Davidson, whose arguments in its favour I will consider in the following two sections:

> nothing ... could count as evidence that some form of activity could not be interpreted in our language that was not at the same time evidence that that form of activity was not speech behaviour. (1984, p. 185)

As a point against untranslatability the argument is this. For neither an untranslatable language nor for non-linguistic behaviour can semantic content be given in our language. Thus inability to translate is indeterminate between being evidence that a language is untranslatable and that it is not a language at all.

Now it is true that attempted translation of an untranslatable language and of non-linguistic behaviour both result in translation failure. But, as Davidson himself notes, to conclude from this that no evidence could show an untranslatable language to be a language "comes to little more than making translatability into a familiar tongue a criterion of languagehood" (1984, p. 186). It is not just that it makes translatability criterial for languagehood in the unobjectionable sense of being a sufficient condition. It takes it as a necessary condition. But translation into our language cannot be necessary for identifying a language as such. As pointed out in 4.3, there are translation-independent ways of recognizing language.

Davidson, however, denies that language is conceivable or recognizable as such independently of translation. The central argument of his (1984) is directed against "the dualism of conceptual scheme and empirical content" which underlies the conception of language as independent of translation. This "dualism" posits an opposition between language, which embodies a conceptual system, and reality, upon which that system imposes order. The opposition of scheme versus content bypasses translation and characterizes language as a relation between thought and reality. Thus a language is something that bears the scheme-content relation to reality. Because this characterization of language dissociates it from translation, Davidson must dispose of the dualism to press home the argument that evidence for an untranslatable language is indeterminate.

Before discussing Davidson's attack on the dualism a caveat must be lodged: the relevance of the attack to incommensurability is open to question. Davidson equates incommensurability with total translation failure and attacks the scheme-content dualism as underlying the idea of such total failure. He treats total and partial failure as separate issues, deploying the principle of charity against the latter. The problem is that while partial is distinguished from total translation failure, nothing is said about where languages begin and end or about the possibility of nested languages. But if total translation failure is a relation between total natural languages, since the dualism is connected with total failure it may be deemed irrelevant to untranslatability between theoretical sub-languages. On the other hand, it is not immediately obvious that the relation between theoretical sub-languages fails to be relevantly similar to the relation between total languages. (Why not consider such sub-languages the embodiment of total conceptual schemes?) But even though the relevance of the attack on the dualism is thus unclear, it is instructive to view Davidson's arguments in the light of the present analysis of incommensurability.

According to Davidson, the scheme-content dualism disconnects languagehood from translation as follows:

something is a language, and associated with a conceptual scheme, whether we can translate it or not, if it stands in a certain relation (predicting, organizing, facing, or fitting) to experience (nature, reality, sensory promptings) ... The images and metaphors fall into two main groups: conceptual schemes (languages) either *organize* something, or they *fit* it (as in 'he warps his scientific heritage to fit his ... sensory promptings'). The first group contains also *systematize*, *divide up* (the stream of experience); further examples of the second group are *predict*, *account for*, *face* (the tribunal of experience). As for the entities that get organized, or which the scheme must fit ... either it is reality (the universe, the world, nature), or it is experience (the passing show, surface irritations, sensory promptings, sense-data, the given). (1984, pp. 191-2)

Such a relation gives substance to languagehood not contingent upon an interlinguistic relation of translatability. Identification of a language need not therefore require translation, but may proceed via evidence of the right sort of relation between putative linguistic behaviour and the world. Thus the dualism allows that a language might be recognized as such without translation into a home language.

To continue the theme of the caveat, if a conceptual scheme is a total language, the dualism is ill-suited to discussion of incommensurability. It is true that untranslatability between theories is due to difference in concepts applied to a common domain. To that extent, the dualism relevantly applies to untranslatability between theories. However, if recognition of an instance of the scheme-content relation is meant to tell us when we are dealing with language at all, as opposed to non-linguistic behaviour, then it is irrelevant. For questions of language possession do not arise, or are already settled, between adherents of incommensurable theories. Since untranslatable theories may be set within a shared background language, interpretation of a rival theorist is not an exercise in radical translation (see 4.3).

On Davidson's analysis, the scheme-content relation can be parsed in several ways, depending on what schemes do and what they do it to. There are two pairs of relations: either schemes organize reality or experience, or they fit reality or experience. The accent with the first pair is on the taxonomic role language plays in dividing the world into reference classes. With the second it is on predictive or explanatory success. Against each pair Davidson develops one main line of argument. He argues that the organizing idea does not give content independent of translatability to the idea of a language. The idea of language fitting reality, on the other hand, separates truth from translation and leads illegitimately to the idea of a true but untranslatable language.

4.5 Schemes organize the world

The organizing idea is that something which is a language is recognizable as such because of its classificatory function. Translation fails because languages arrange things differently.

Against this version of the dualism, Davidson first notes that only pluralities can be organized:

> We cannot attach a clear meaning to the notion of organizing a single object (the world, nature etc.) unless that object is understood to contain or consist in other objects. Someone who sets out to organize a closet arranges the things in it. (1984, p. 192)

He then argues that it can only be determined that a language organizes things differently if the language can on the whole be translated:

> A language may contain simple predicates whose extensions are matched by no simple predicates, or even by any predicates at all, in some other language. What enables us to make this point in particular cases is an ontology common to the two languages, with concepts that individuate the same objects. We can be clear about breakdowns in translation when they are local enough, for a background of generally successful translation provides what is needed to make the failures intelligible. But we were after larger game: we wanted to make sense of there being a language we could not translate at all. Or, to put the point differently, we were looking for a criterion of languagehood that did not depend on, or entail, translatability into a familiar idiom. I suggest that the image of organizing the closet of nature will not supply such a criterion. (1984, p. 192)

So, while admitting extensional variance between languages, Davidson denies that "organizing the closet of nature" gives translation-independent content to languagehood. The crux of the argument is the assumption that without translation we could not tell that a language divides the world up differently, or indeed that it divides the world up at all.

It is unclear just how the point that only a plurality of items can be organized bears on the argument. Perhaps it supports the claim that languages which divide up the same reality have a common ontology of objects. But this claim is inapplicable to theories: theories in the same domain may quantify over different theoretical entities. Or perhaps the point is that reality divides itself up into a set of objects before language organizes it. So languages all organize the same plurality of objects. Still, they might group the same objects differently, with translation failure as a result. In any case, why should a fixed plurality exist prior to the imposition of language? The fact remains that a unity cannot be organized even if languages create different objects. Of course, the point may simply be a concession of cross-language semantic diversity. Davidson can well grant that languages range over pluralities and admit extensional variance across languages. Having conceded this much, he may still deny that the organizing idea sustains total translation breakdown.

Extensional variance raises the possibility of translation failure. Davidson's tactic is to play down its scope. Rather than argue against semantic differences between languages, he argues that there are limits to the intelligibility of such difference. Translation failure, he says, must be limited if it is to be intelligible: "we can be clear about breakdowns in translation when they are local enough". The point is directed against the possibility of a total translation failure, of which he thinks no sense can be made. As such, however, the point is no objection to incommensurability. Intertheoretic untranslatability constitutes at most local translation failure. As sub-languages, the terminologies of theories are embedded in global natural languages. So, in full accord with Davidson, translation between theories may very well fail against "a background of generally successful [albeit homophonic] translation".

As noted above, the crux of the argument is that it is necessary to translate in order to determine that a language divides the world up differently. Within the argument this assumption serves as support for the conclusion that translation failure is intelligible only if it is local and is set within the context of broad translational success. Of course, if either the assumption or the conclusion were true, then the idea of organizing reality would not offer a means independent of translation for recognizing a language.

But, given the aim of the argument, the assumption begs just the question at issue. Granted, it seems true that in order to find out that the classificatory systems of languages differ, the languages must be understood.[11] But Davidson simply assumes that translation is necessary for the understanding of another language and of the classificatory system it embodies. To allow such an assumption is to lose sight of the purpose of the argument. Davidson is in the process of arguing that the idea of organizing reality does not give content which is independent of translation to the notion of being a language. In order to show this, it has first to be argued that there is no way to determine whether a language organizes reality without translation. But this Davidson simply assumes. Surely, in the context of arguing that translation is necessary for the determination of classificatory difference, it begs the question to assume that translation is the only way to find out about such difference.

Davidson's conclusion is that the "image of organizing the closet of nature" does not enable sense to be made of total translation failure. This conclusion bespeaks a certain verificationism.[12] For it assumes that failure to specify a test for the presence of an untranslatable language entails that no content has been given to the concept of such a language. And to assume that concepts only have content if there is a test for their application is to assume that meaning consists in verification.

Davidson's verificationism is evident in his inference from the intelligibility of only local translation failure to the unintelligibility of total failure. He allows that "we can be clear about breakdowns in translation when they are local enough". And he claims that general success in translation is what makes such breakdowns "intelligible". Davidson's point is that untranslatable linguistic material can only be known to be language (as against not being linguistic) if translation failure is local in the sense that it occurs in the context of overall translation of the language; otherwise there would be no semantic evidence that the untranslatable material is linguistic. From this Davidson infers that sense has failed to be made of total failure: "But we were after larger game: we wanted to make sense of there being a language we could not translate at all."

The inference appears to be based on the following reasoning. The idea that language organizes reality can only be applied to the local case given background success in translating the language. Therefore the organizing idea does not yield a test for determining the presence of a totally untranslatable language. So that idea does not give meaning to the concept of such a language. This final inference assumes that meaning is only bestowed upon concepts if a means of verification is specified.

The problem with this can best be seen from Davidson's (previously quoted) alternative formulation of his conclusion:

> ...to put the point differently, we were looking for a criterion of languagehood that did not depend on, or entail, translatability into a familiar idiom. I suggest that the image of organizing the closet of nature will not supply such a criterion. (1984, p. 192)

By 'criterion' Davidson seems to mean a test for language, not an account of what being a language consists in. He concludes that because the criterion is inapplicable without translation no sense has been made of full untranslatability. But the "image of organizing the closet of nature" specifies a function which a language may perform. In that respect, it offers a criterion of being a language, as opposed to a criterion for recognizing one.[13] It thereby gives some content to the notion of being an untranslatable language: viz. such a language organizes the world differently. Such a criterion of languagehood gives content to the notion of being an untranslatable language whether or not it can verifiably be fulfilled.

Such verificationism is objectionable because it imposes a fallacious constraint on meaning. To impart meaning to a concept cannot be contingent upon coming up with a test for applying it. For it is possible to specify mistaken tests for applying concepts. What enables this point to be made with respect to particular concepts is a grasp of their content which is independent of such tests.

Even if this were not the case, Davidson's attack would still be beside the point. For to argue that we do not understand the notion of total untranslatability does not itself constitute an attack on the idea of an untranslatable language. Even if we have no conception of what a fully untranslatable language would involve, no existence claim follows from that about such languages. Clearly what can be conceptualized on the basis of present linguistic resources is neither to be taken as exhaustive of what actually exists nor as somehow imposing any limits on existence. Neither from inability to verify the existence of a totally untranslatable language, nor from inability to give content to the concept of such a language, does it follow that no such language exists.

4.6 Schemes fit experience or reality

Our next topic is Davidson's criticism of the idea that conceptual schemes fit the world or experience. He argues that this construal of what schemes do effects an untenable separation between truth and translation.

On this version of the dualism, a conceptual scheme or a language enables us to deal with the world by explaining and predicting facts: schemes are a way of "coping with (or fitting or facing) experience" (p. 193). Such metaphors differ from the organizational image in that they emphasize prediction over classification. Hence, they take us away "from the referential apparatus of language ... to whole sentences":

> It is sentences that predict (or are used to predict), sentences that cope or deal with things, that fit our sensory promptings, that can be compared or confronted with the evidence. (1984, p. 193)

The relation between the two versions of the dualism appears to be this: schemes which organize the world differently provide alternative ways of coping with experience. Since it is "sentences that cope" and the "referential apparatus" from which sentences are built varies with scheme, sentences from alternative schemes may be untranslatable and yet deal adequately with the world.

Davidson first argues that the idea of fitting experience reduces to that of being true. Schemes account for all the evidence:

> ... a theory may be borne out by the available evidence and yet be false. But what is in view here is not just actually available evidence; it is the totality of possible sensory evidence past, present, and future. (1984, p. 193)

To deal with all such evidence is just to be true: "for a theory to fit or face up to the totality of possible sensory evidence is for that theory to be true". There is no need to maintain a dichotomy between fitting all the evidence and being true:

> ... the notion of fitting the totality of experience, like the notion of fitting the facts, or of being true to the facts, adds

> nothing intelligible to the simple concept of being true. (1984, pp. 193-4)

Instead of two versions of what schemes fit we have this: "something is an acceptable conceptual scheme or theory if it is true". Since fitting experience or reality thus reduces to being true, "the criterion of a conceptual scheme different from our own now becomes: largely true but not translatable".

This raises the question of whether "we understand the notion of truth, as applied to language, independent of the notion of translation". Davidson takes Tarski's theory of truth as constitutive of our understanding of truth. Convention T requires translation from an object-language into the meta-language in which the truth-predicate is defined, so our understanding of truth depends crucially on translation. It is worth quoting his remarks in full:

> We recognize sentences like '"Snow is white" is true if and only if snow is white' to be trivially true. Yet the totality of such English sentences uniquely determines the extension of the concept of truth for English. Tarski generalized this observation and made it a test of theories of truth: according to Tarski's Convention T, a satisfactory theory of truth for a language L must entail, for every sentence s of L, a theorem of the form 's is true if and only if p' where 's' is replaced by a description of s and 'p' by s itself if L is English, and by a translation of s into English if L is not English. This isn't, of course, a definition of truth, and it doesn't hint that there is a single definition or theory that applies to languages generally. Nevertheless, Convention T suggests, though it cannot state, an important feature common to all the specialized concepts of truth. It succeeds in doing this by making essential use of the notion of translation into a language we know. Since Convention T embodies our best intuition as to how the concept of truth is used, there does not seem to be much hope for a test that a conceptual scheme is radically different from ours if that test depends on the assumption that we can divorce the notion of truth from that of translation. (1984, pp. 194-5)

So the overall structure of Davidson's attack on the idea that schemes fit experience or reality is a two step argument. The first step is the reduction of the idea to that of being true. The second is the argument that truth is inextricable from translation. The two steps are linked in that the idea of an untranslatable scheme being true divorces truth from translation. Both steps may be criticized.

The problem with the first part of the argument is that fitting experience does not reduce to being true as far as scientific theories are concerned. In science, excessive charity is inappropriate, since theories may be, and often are, mistaken. More to the point, theories which "fit the evidence" in the sense of being empirically adequate may yet be false; for a false theory may entail true predictions.

Davidson does, it is true, restrict attention to theories which fit "the totality of possible sensory evidence past, present, and future". But this simply removes actual science from the ambit of the argument. What he says can neither be about actual science nor is his argument relevant to examples that have been put forward of untranslatable theories. For rarely, if ever, do actual theories fit all the evidence, much less all the future evidence.

Certainly, there is no need to assume purportedly untranslatable theories to be true. To take but one example, the phlogiston theory and Lavoisier's oxygen theory were both to varying degrees false. To say that a pair of theories is incommensurable carries no commitment to their truth: it is not to say that they are both untranslatable and true.

In any event, no realist would grant that a theory which fits all "possible sensory evidence" is ipso facto true. It might be granted that a theory which fits all the facts, observable and otherwise, is true; but if it fits only the "sensory evidence", it does not follow that it is true. Even if a pair of untranslatable theories were to fit all the evidence, there would be no reason to suppose both were true: to describe such a pair as incommensurable is not therefore to say that they are true and untranslatable.

Part of the problem lies with the choice of metaphor. "Fitting the evidence" suggests empirical adequacy, which amounts to truth at an empirical level. But in any sense in which theories "cope with experience" they need not strictly "fit the evidence". Even successful theories in actual science fit the evidence only imperfectly. Theories are beset with empirical difficulties from the outset and no theory ever fits all the evidence. This does not keep them from "coping". Since theories are not directly refutable, counter-instances may be deflected upon auxiliary assumptions; and progress may be made in spite of anomalies. They may still "cope with experience" in the sense of explaining and predicting phenomena, solving problems, and guiding research. To say that such theories "fit the evidence" in any but a loose sense is mistaken. There is even less reason to say that they are true.

Thus fitting experience does not reduce to being true. Theories may cope with experience without full empirical adequacy and they may fit the evidence without being true. Pairs of untranslatable theories may be partly or wholly false. Thus there is no commitment to incommensurable theories being both true and untranslatable. Untranslatability is due to semantic differences which obtain whether theories are true or false.

Since incommensurability is not necessarily a relation between true theories, it is tempting to think that this removes the need to make sense of truth independently of translation and hence breaks the link between the two steps of Davidson's argument. However, Davidson's Tarskian argument cannot be evaded so easily. For the idea of untranslatability itself implies the possibility of true but untranslatable sentences. This can be seen as follows. If a sentence can be formulated in a language, then under ordinary circumstances either it or its negation is true. If a sentence cannot be translated from one language into another, then neither can its negation be so translated. Since either the sentence or its negation is true, untranslatability raises the possibility of a true but untranslatable sentence. So Davidson's attack on the separation of truth from translation must be confronted.

Davidson argues that our concept of truth is defined for English and languages translatable into English, so our grasp of the concept does not extend beyond languages intertranslatable with English. Convention T does not define a general concept of truth for unspecified languages. Rather, it defines a truth-predicate for a specific language (in our case English) and for sentences of languages intertranslatable with it.

A theory of truth for a language which conforms with Convention T entails a set of T-sentences for the sentences of the language and their translational equivalents. Recurring to the previous quotation:

> ...according to Tarski's Convention T, a satisfactory theory of truth for a language L must entail, for every sentence s of L, a theorem of the form 's is true if and only if p' where 's' is replaced by a description of s and 'p' by s itself if L is

English, and by a translation of s into English if L is not English. (1984, p. 194)

The set of English T-sentences defines the English truth-predicate for the sentences of English and translational equivalents:

...sentences like '"Snow is white" is true if and only if snow is white' [are] trivially true... the totality of such English sentences uniquely determines the extension of the concept of truth for English. (1984, p. 194)

Since no T-sentence can be formed in English for sentences not translatable into English, the truth-predicate of English is not defined for such sentences, which therefore fall outside its extension.

Thus our concept of truth is given by the definition of the English truth-predicate which is defined exclusively for the set of English sentences and translational equivalents. Such a concept of truth cannot be understood independently of translation. For it would not be constitutive of understanding that concept to understand it as applied to untranslatable sentences: it would not be that concept if so applied.

On the face of it, this argument seriously undermines the idea that there might be a total natural language which is completely untranslatable into another. For if something is a language then it should be possible to formulate a true sentence in it. But if such a language were totally untranslatable we would be unable to attach any sense to the idea of its sentences being true. That is, using our truth-predicate, we would be unable to say of any such untranslatable sentences that they were true.

As will be argued subsequently, this argument does not affect the possibility of translation failure between parts of a single language. So long as such sub-languages are contained within a more inclusive language, the truth-predicate may be defined over them in the containing language. Hence Davidson's argument poses no threat to the thesis of untranslatability of theoretical terminologies embedded in a background language. The argument may nonetheless be criticized on several points.

In the first place, the argument does not achieve its aim. It is meant to show, as against the scheme-content dualism, that something crucial to being a language (true assertion) has no content divorced from translation. But in order to show that one could not discover a language which turned out not to be translatable, it needs to be shown that a language could not be recognized as such without translation. What it purports to show is instead that truth is indefinable for untranslatable sentences. But that does not show that a language could not be identified as such from non-semantic evidence. If a language which proved resistant to translation were to be so identified, that would present a posteriori the existence of untranslatable truth. In denying that truth can be disjoined from translation, Davidson rules out untranslatable truth a priori. But no argument is offered from the connection between truth and translation to the conclusion that language is unrecognizable as such in the absence of translation. So far from showing the impossibility of such language recognition, the argument merely assumes it.

In the second place, there is an underlying tension between the purported truth-translation nexus and Davidson's concession, noted in 4.5, of local translation failure. As we saw, Davidson allows that "we can be clear about breakdowns in translation when they are local enough" (p. 192). But if a sentence of a language which is on the whole translatable into English should turn out not to be so translatable, what is to be made of the possibility of its truth?

According to Davidson, the English truth-predicate is undefined for any sentence untranslatable into English. So on Davidson's own account our concept of truth is inapplicable to such a sentence. Yet either such a sentence or its denial is true. Whatever sense Davidson thinks can be made of the idea of an untranslatable sentence, he seems not to allow sense to be made of its truth.

Now such isolated translation failure might be dismissed as merely pragmatic and hence unproblematic. Languages evolve divergently and linguistic modifications may remove local untranslatability. The fact that truth-conditions cannot be given for isolated sentences need not preclude sense being made of their truth. For, suitably modified, the language may translate recalcitrant sentences and subsume them under its truth-definition.

But when does it become intelligible to apply the concept of truth to such a sentence? If the sentence must await actual translation, problems arise with translating the truth-predicate. For until such an untranslatable sentence can be translated, the truth-predicate defined in its language does not have the same extension as ours. If our concept of truth can be applied to such a sentence prior to the requisite alteration of our language, then our truth-predicate can be applied to sentences for which no T-sentence in our language can be formed.

In any case, to translate by altering a language is not strictly translation at all. If a sentence may only be translated by changing a language, then it cannot be translated into the unchanged language. But linguistic boundaries are fluid and arbitrary. No rules dictate when a fragment of a language becomes part of another or how large such a fragment may be. In principle, nothing prevents one language being appended in its entirety onto another. To permit application of the truth-predicate to sentences translatable by linguistic modification amounts to making the possession of truth-value depend on whether a sentence belongs to our language. But to have a truth-value is not merely contingent upon belonging to our language. Nor does a sentence acquire truth-conditions only upon entry into our language.

In the third place, at least a prima facie case can be made that truth is separable from translation. Suppose one were to protest against Davidson that the concept of truth does not depend on translation. The Tarskian schema "'s' is true if and only if p" supplies a structural feature of truth which does not merely consist in a specification of the extension of 'true' for English. It is a constraint on the concept of truth such that nothing counts as a truth-predicate unless the sentence of which truth is predicated and the statement of truth conditions are equivalent. As against Davidson, the suggestion is that there is a general concept of truth of which the truth-predicates of particular languages are special cases.

To give some content to this claim, let us consider how one might come to recognize a truth-predicate for an untranslatable language. Consider a field linguist whom we may imagine to have encountered and mastered an alien language, call it "Alien", which fails to translate into the linguist's home language, say English. What is to prevent such a linguist from recognizing an Alien predicate whose use in Alien corresponds to the behaviour in English of the predicate 'is true'? Suppose the linguist identifies a predicate 'T' of Alien such that appending 'T' to a named Alien sentence 's' yields a sentence "'s' is T" which is assertible when and only when 's' is assertible. Provided the linguist understands what 's' means and understands that "'s' is T" is materially equivalent to 's', what reason could there be not to take 'T' as the truth-predicate for Alien?

Davidson's argument suggests the following objection to this proposal. Suppose the linguist reports in English, as regards the Alien sentence 's', that 's' is true. What does the linguist's report "'s' is true" mean? Since truth-conditions cannot be given for 's' in English, the English truth-predicate cannot be used to say that 's' is true. So to say in English that 's' is true must mean that 's' is true-in-Alien, not true-in-English. But what does "'s' is true-in-Alien" mean in English? 'True-in-Alien' is indefinable in English because no Alien truth-conditions are specifiable in English.

To give sense to saying "'s' is true-in-Alien" in English one might say that 'true-in-Alien' is English for the Alien truth-predicate. 'True-in-Alien' and 'true-in-English' have similar functions in their respective languages. Each predicate behaves disquotationally: the result of appending either predicate to a sentence is a sentence assertible in identical circumstances to the original. In virtue of this formal resemblance both predicates instantiate a general truth-concept for particular languages, and 'true-in-Alien' can be used in English to translate the Alien truth-predicate.

It may be objected that the notion of 'true-in-Alien' is inconsistent with a semantic conception of truth. Since no truth-condition can be given for 's' in English, what it is to say "'s' is true-in-Alien" in English cannot be defined in English.

Now, we may grant that the extension of the truth-predicate for a language is defined within the language by its T-sentences. No extensional specification of 'true-in-Alien' can be given in English using English T-sentences since Alien is untranslatable into English. But it does not follow that no content can be given to 'true-in-Alien' in English. For the fact that the function of the Alien predicate is analogous to that of English 'true' enables 'true-in-Alien' to be defined as an English word for the Alien predicate which performs the same function in Alien as 'true' does in English.

It might be further objected that the Alien truth-predicate is not recognizable as such if it differs extensionally from English 'true'. It is not in virtue of disquotation that a truth-predicate is identifiable as such. In order to identify a truth-predicate, its extension must be determined. To identify such an extension as the extension of a truth-predicate, it must be the same extension as the extension of the English truth-predicate.

As against this, the way our imagined linguist recognizes the Alien truth-predicate is precisely the same way in which the truth-predicate for English is identified. Given that the linguist understands Alien and recognizes a predicate of Alien whose behaviour conforms to Tarski's schema, nothing further is required for recognizing a truth-predicate. The objection reduces, in effect, to the denial that understanding a language is possible without translation. This denial was found questionable in 4.3.

Fortunately, translation failure between theories does not require a translation-independent concept of truth. As argued in 4.2, the languages of scientific theories constitute sub-languages embedded within a background natural language. Such theoretical sub-languages may be discussed within the inclusive natural language employed as a metalanguage. This point enabled us to meet the direct incoherence argument by saying that the language in which untranslatability is argued for and the language into which translatability fails are not the same.

A related point applies here. Since English may function as a metalanguage, the English truth-predicate can be defined over its embedded sub-languages. In particular, English may be employed as a metalanguage in which to define truth over the sub-languages of theories. So, for example, English T-sentences for sentences of the impetus theory and of Newtonian mechanics may be formulated as follows:

> 'Projectile bodies have impetus' is true if and only if projectile bodies have impetus.
>
> 'Projectile bodies have momentum' is true if and only if projectile bodies have momentum.

Since English contains the sub-languages of both theories there is no need to characterize the truth in English for sentences not translatable into English. The fact that the terminology of the impetus theory cannot be translated into that of Newtonian mechanics poses no special problem. Provided that such terminologies are in fact sub-languages of English, English may function as a metalanguage and the English truth-predicate may be defined for both. For it is not translation into English that is denied. Rather, what is denied is that expressions of the impetus theory can be translated into the language of Newtonian mechanics. That is untranslatability within English, not into English.

In sum, Davidson's argument poses a problem for the idea of a language altogether untranslatable into ours. The argument may be met if content can be given to a general concept of truth not defined by giving its extension for English sentences. However, no analogous problem arises for the thesis that theories within a language may be untranslatable. For such a thesis does not depend on a generalized conception of truth and is consistent with a notion of truth defined for English.

Notes

1. Of course, the criticism can be extended to arguments which do not use examples because the use of examples in independent contexts is inconsistent with such arguments. But that does not refute the arguments; it only shows it is inconsistent to use examples and arguments.
2. See Kuhn's discussion of Newton and Einstein (1970a, pp. 101-2), phlogiston versus oxygen (1983, pp. 675-6); Feyerabend on impetus and momentum (1981d, pp. 62-9), and classical physics versus general relativity and quantum mechanics (1981e).
3. Broadening the extension of 'incommensurable' was an afterthought. Feyerabend's original discussions concern scientific theories, and it is only with his (1975, ch. 17) that he actually applies it to other "structures of thought". Claims of generality are all recent (e.g. 1975, p. 269, 1981b, p. 16, fn. 38, 1987, p. 81).
4. E.g. "... the conditions of concept formation in one theory forbid the formation of the basic concepts of the other" (1978, p. 69, fn. 118); cf. (1987, p. 81).
5. Though they deny translation between theories, Kuhn and Feyerabend both allow some semantic parallels. Feyerabend concedes that "incommensurable concepts may exhibit many structural similarities" (1975, p. 277). Kuhn concedes co-reference of tokens (1983) and grants that "translation of one theory into the language of another depends ... upon compromises ... whence incommensurability" (1976, p.191).
6. For further defence of the distinction between translation and understanding, as well as elaboration of some of the other themes of this section, see my (1991c).
7. In fact, as I argue in my (1992), proof of languagehood must ultimately rest on such non-semantic factors.
8. I say "derives" advisedly. Davidson puts the principle of charity to a different use. However, the argument discussed in the text follows immediately from Davidson's analysis of

interpretative charity. He uses the principle against partial translation failure and concludes that no sharp distinction between difference of language and of belief can be drawn.

9. The point is made by Newton-Smith (1981, p. 163).
10. E.g. the problem of induction, underdetermination of theory by data, theory-ladenness of observation, etc.
11. Even this is unproven. It is not a priori true that the only way to determine classificatory difference is by analysis of content.
12. A number of authors have noted Davidson's implicit verificationism here: among them Rorty (1982, pp. 5-6) and Blackburn (1984, p. 61).
13. Clearly, it cannot be a sufficient condition, but it is perhaps necessary.

5 Referential discontinuity

5.1 Introduction

This chapter is concerned with the idea that the transition between incommensurable theories involves significant change of reference. The idea occurs in both radical and moderate forms. In its radical form, the idea is that incommensurable theories have little or no reference in common, so that the transition between them involves wholesale discontinuity of reference. The radical reference change thesis is found in Feyerabend, as well as in Kuhn's original treatment of the subject. In Kuhn's later writings the thesis of reference change takes a moderate form, on which such change is restricted to theoretically central terms.

This chapter extends the discussion of reference in Chapter Two by considering in detail what Kuhn and Feyerabend actually say about reference change. It differs from the earlier discussion in that it presents a close analysis of their treatment of the relation between conceptual change and reference change. The aim of the chapter is to bring out the difficulties which confront Kuhn's and Feyerabend's treatment of reference in the light of our earlier discussion of meaning and reference.

In Chapters Six and Seven extreme versions of the reference change thesis will be dealt with. Chapter Six shows that the extreme idealist thesis that the world itself literally changes cannot be attributed to Kuhn and Feyerabend. Chapter Seven criticizes less extreme views according to which the world of a theory is not the real world, but is in some way constituted by theory. This chapter concerns the least extreme reference change view which allows reference to change profoundly while holding the world constant.

The chapter is organized as follows. In 5.2 Feyerabend's position on radical reference change is interpreted and criticized. Section 5.3 concerns Kuhn's original position, which is shown to be ambiguous. Section 5.4 treats Kuhn's later, more moderate reference change thesis.

5.2 Feyerabend

Feyerabend is committed to two theses which are in tension with each other. His view of reference commits him to discontinuity of reference between incommensurable theories. His account of observation commits him to continuity of reference to observable objects.

Feyerabend employs a pragmatic account of observation. According to this account, observation statements are observational by virtue of the physical and causal circumstances in which they are used rather than because of their meaning. As he explains:

> I shall call this account the pragmatic theory of observation. The theory admits that observational sentences assume a special position. However, it puts the distinctive property where it belongs, viz., into the domain of psychology: observational statements are distinguished from other statements not by their meaning, but by the circumstances of their production. (1965, p. 212)

As opposed to a "semantic theory of observation", which "assumes that terms are observational by virtue of their meanings" (1965, p. 198), a pragmatic account provides a non-semantic analysis of observation statements:

> a statement will be regarded as observational because of the causal context in which it is being uttered, and not because of what it means. According to this theory, "this is red" is an observation sentence because a well-conditioned individual who is prompted in the appropriate manner in front of an object that has certain physical properties will respond without hesitation with "this is red"; and this response will

> occur independently of the interpretation he may connect with the statement [he may interpret it as referring to a property of the surface of the object, as a property of the space between the object and the eye (as did Plato), as a relation between the object and a coordinate system in which he himself is at rest]. (1965, p. 198)

On this account, an observation statement is characterized as such by the circumstances in which it is used. That is, an observation statement is distinguished by its standard application to states of affairs which are so situated with respect to an observer that the production of the statement is the result of the observer's perception of such a state of affairs.

As characterized by Feyerabend, the procedure which leads to the application of an observation sentence is independent of meaning. Although he holds that the meaning of observational statements is determined by theory, he allows that the application of observational statements may in fact be invariant relative to change of meaning. But it follows from the invariance of the pragmatic conditions of the use of an observational sentence that the objects to which the sentence is applied must be invariant as well.[1]

For consider the second of the above passages. Feyerabend notes that an observer may describe an observed object as red independently of the meaning of the term 'red'. That is, the pragmatic application conditions of the observation statement 'This is red' need not be affected by variation in the meaning of the sentence. But that implies that the application of the sentence to the same set of red individuals is constant throughout such change of meaning.

Feyerabend is explicit that the pragmatic conditions in which observation sentences are applied may remain invariant while the meaning of such sentences is changed:

> general ideas may change without any corresponding change of observational procedures. For example, we may change our ideas about the nature, or the ontological status (property, relation, object, process, etc.) of the color of a self-luminescent object without changing the methods used for ascertaining that color (looking, for example). Clearly, such a change is bound profoundly to influence the meanings of our observational terms. (1965, p. 170)

> the scientist who employs parts of everyday language for the purpose of giving an account of his experiments does not introduce a new use for familiar words such as 'pointer', 'red', 'moving', etc. whenever he changes his theories. (1981c, p. 31)

Though the use of observational language may survive variation of meaning, such use need not be completely invariant; for, as Feyerabend notes, use may change too (1981c, p. 31).

It follows from the possibility of invariant application conditions through variation of meaning that at least a modicum of extensional invariance and intersection can obtain between incommensurable theories. For consider the sentence 'This is red'. Because the application conditions of the sentence are invariant, it continues to be applied to the same set of red individuals even though its meaning changes. This implies that the extension of 'red' is at least partially preserved throughout shift of meaning. Even if the extension of 'red' changes from a property to a relation, the same set of red objects still belongs to the extension.

More generally, the stability of the use of observational language implies that incommensurable theories may be applied to the same observable entities. Hence reference to a common set of observable objects may be preserved in the transition between such theories. At the very least, there may be extensional overlap at the observational level.

The problem is that continuity of reference at the observational level is inconsistent with discontinuity of reference between incommensurable theories. It will now be shown that Feyerabend is committed to referential discontinuity, so the conflict is genuine.

In Feyerabend's view, neither the pragmatic conditions of an observation statement nor the phenomenological experience which accompanies it determine its meaning. Rather, its meaning depends on theoretical context.[2] That is, the meaning of observational terminology is determined by a theory which explains the entities to which the terms are applied. Thus the meaning of the term 'red' is not given by the pragmatic features of its use nor by the sensation of the colour red. It is given by a theory of redness and may vary with change of theory.[3] Thus, retained vocabulary may undergo change of meaning in the transition between theories, and new terms with new meanings may be introduced. The meaning variance that results is the basis of the idea of incommensurability.

I will now examine Feyerabend's argument that the concept of impetus is incommensurable with the concepts of Newtonian physics (1981d, pp. 65-7). The argument is divided into two sub-arguments. The first argument is that the concept of impetus cannot be defined on the basis of Newtonian concepts. The second is that 'impetus' cannot be brought into a relation of co-extensiveness with Newtonian expressions. From such failure of either analytic or extensional relations to obtain, Feyerabend concludes that the theories are incommensurable.[4]

Let us consider the first part of Feyerabend's argument, which is that the concept of impetus is indefinable within Newton's physics. The argument depends on the incompatibility of the definition of impetus with basic Newtonian principles. The concept of impetus presupposes that all sustained motion requires continuous causation. Impetus was thought to be a force imparted to a body by an external action, which sustains the body's motion by acting upon the body even after contact with the external force ceases. But the presupposition of the concept was denied by Newton, according to whom sustained inertial motion does not require continuous causal influence.

Feyerabend notes, first of all, that 'impetus' is not synonymous with 'momentum': "whereas the impetus is supposed to be something that pushes the body along, the momentum is the result rather than the cause of its motion" (1981d, p. 65). Then he points out that the very existence of impetus is precluded by the Newtonian analysis of inertial motion: "the inertial motion of classical mechanics is a motion which is supposed to occur by itself, and without the influence of any causes" (p. 65). Finally, he argues that the concept of impetus cannot be defined within the Newtonian framework: "given that the movement under review (the inertial movement) occurs with constant velocity, and Newton's second law, we obtain in all relevant cases zero for the value of the force, which is not the measure we want" (p. 66). Since the concept of impetus depends on the principle that all motion requires a continuous sustaining cause, Feyerabend concludes that it "involves laws ... which are inconsistent with Newtonian physics" and "cannot be defined in a reasonable way within Newton's theory" (p. 66).

This argument shows that 'impetus' cannot be brought into an analytic relation with Newtonian concepts, so is essentially an argument about meaning. However, in an indirect way it is about reference as well. For inability to define 'impetus' within the Newtonian framework stems from the denial of the existence of a force which sustains inertial motion. It follows from such a denial that the term 'impetus' fails to refer from a Newtonian point of view.

This suggests that the indefinability of such a concept is closely connected with failure of co-reference. But while the connection is only implicit in the first part of the argument, in the second part of the argument it becomes explicit. For Feyerabend simply transfers the above considerations about meaning to the issue of reference.

The second part of Feyerabend's argument is directed against the idea that 'impetus' can be related to the Newtonian conceptual framework by a non-definitional connection. Nagel had suggested that, in the place of analytic relations between the concepts of theories, empirical correlations might obtain between such concepts. According to Nagel, such concepts might be linked by means of

> a material, or physical hypothesis according to which the occurrence of the properties designated by some expression in the premises of the primary science is a sufficient, or a necessary and sufficient, condition for the occurrence of the properties designated by the expressions of the secondary discipline.[5]

Such an interrelation would be empirical because in the absence of analytic connections between concepts, the possibility of co-reference remains. Given that such a relationship is not a relation of meaning, it would obtain in virtue of empirical facts about reference. In the following passage, Feyerabend argues that this second approach cannot be applied to the case of impetus either:

> this method amounts to introducing a hypothesis of the form
>
> impetus = momentum
>
> where each side retains the meaning it possesses in its respective discipline. The hypothesis then simply asserts that wherever momentum is present, impetus will also be present (see the above quotation of Nagel's), and it also asserts that the measure will be the same in both cases. Now this hypothesis, although acceptable within the impetus theory (after all, this theory permits the incorporation of the concept of momentum), is incompatible with Newton's theory. It is therefore not possible to achieve reduction and explanation by the second method. (1981d, p. 67)

The hypothesis is incompatible with Newtonian theory because the latter precludes the existence of a force acting upon inertial motion. Feyerabend takes this to show that 'impetus' cannot be empirically correlated with any Newtonian concept.

The proposed correlation is equivalent to the assertion that 'impetus' and 'momentum' have the same extension. So the argument is meant to show that the hypothesis is an incorrect assertion of co-reference. The inference that the correlation fails to obtain must therefore depend on an assumption about reference.

The reason Feyerabend gives for the failure of co-extensiveness is that the correlation is "incompatible with Newton's theory". That is, the correlation fails because the existence of impetus is incompatible with Newtonian physics. This assumes that terms whose definitions are jointly inconsistent cannot co-refer. Such an assumption is in turn supported by the fact that inconsistent descriptions cannot be jointly satisfied. But the fact that inconsistent descriptions cannot be true of the same things only entails failure of co-reference on the further assumption that reference is determined by the satisfaction of a description. Thus, Feyerabend's argument against the second way of connecting 'impetus' with Newtonian concepts depends on a description theory of reference.[6]

The difference between these two arguments of Feyerabend's is that the first concerns meaning while the second concerns reference. The conclusions of the two arguments are correspondingly different: no analytic relationship as opposed to no common reference. However, the two arguments are based upon the same consideration. That is, both arguments depend on the incompatiblity of the concept of impetus with basic Newtonian principles. In the case of the first argument, such incompatibility is taken to show absence of analytic connection. In the case of the second, the same incompatibility is taken to show an absence of extensional relation. This parallel between the two arguments is indicative of an underlying assumption that facts about meaning determine facts about reference.

Feyerabend explicitly draws the parallel between the two arguments. In the following summary of his argument, he takes both the absence of analytic relations and the absence of co-reference to follow from considerations about the meaning of 'impetus':

> it was shown that the 'inertial law' [ ] of the impetus theory is incommensurable with Newtonian physics in the sense that the main concept of the former, the concept of impetus, can neither be defined on the basis of the primitive descriptive terms of the latter, nor related to them via a correct empirical statement. The reason for this incommensurability was also exhibited: although [the impetus law of inertia], taken by itself, is in quantitative agreement both with experience and with Newton's theory, the 'rules of usage' to which we must refer in order to explain the meanings of its main descriptive terms contain [the law that motion requires a continuous cause] and, more especially, the law that constant forces bring about constant velocities. Both of these laws are inconsistent with Newton's theory. Seen from the point of view of [Newton's] theory, any concept of a force whose content depends on the two laws [motion requires a continuous cause and the impetus law of inertia] will possess zero magnitude, or zero denotation, and will therefore be incapable of expressing features of existing situations. (1981d, pp. 76-7)

This assumes that it is the relation between the definition of 'impetus' and Newtonian principles that determines the failure of the correlation hypothesis. Since the latter is a hypothesis of co-extensiveness, Feyerabend clearly assumes that considerations about the relations between concepts are capable of deciding questions of co-reference.

Elsewhere, Feyerabend makes this close connection between meaning and reference central to his analysis of incommensurability. In his (1981e) Feyerabend associates change of meaning with discontinuity of reference, effectively unifying the above two patterns of argument.

In the following passage, Feyerabend defines the meaning change of relevance to incommensurability in terms of referential or classificatory change:

> a diagnosis of stability of meaning involves two elements. First, reference is made to rules according to which objects or events are collected into classes. We may say that such rules determine concepts or kinds of objects. Secondly, it is found that the changes brought about by a new point of view occur within the extension of these classes and, therefore, leave the concepts unchanged. Conversely, we shall diagnose a change of meaning either if a new theory entails

> that all concepts of the preceding theory have zero extension or if it introduces rules which cannot be interpreted as attributing specific properties to objects within already existing classes, but which change the system of classes itself. (1981e, p. 98)

On this account the notion of a semantical or linguistic rule is crucial. For it is rules which determine concepts as well as their extensions. And it is change of rules which leads to change of meaning and reference.

Feyerabend does not define the notion of a rule. But it is clear from context that fundamental theoretical laws or principles are meant to count as rules. Moreover, in a closely related context he does comment that:

> [T]he rules (assumptions, postulates) constituting a language (a 'theory' in our terminology) form a hierarchy in the sense that some rules presuppose others without being presupposed by them ... [T]he customary concept of meaning is closely connected, not with definitions which after all work when a large part of a conceptual system is already available, but with the idea of a fundamental rule, or a fundamental law. Changes of fundamental laws are regarded as affecting meanings while changes in the upper layer of our theories are regarded as affecting beliefs only. (1981f, p. 114, fn. 27)

The connection between basic laws and meanings is exemplified by the case of impetus discussed above. For the formulation of the concept of impetus within Newtonian physics is precluded by the law of inertia.

Even without a definition, however, the nature of such rules may be inferred from their function. Feyerabend says in the first of the two preceding passages that "objects or events are collected into classes" by means of rules. Presumably, a rule determines a class by specifying a property as criterial for class membership. So something which lacks a criterial property is excluded from class membership, and a rule which is satisfied by nothing defines an empty class. Such determination of the extension of concepts by rules conforms with the description theory of reference. For according to the description theory, the extension of a term is the set whose members instantiate the property specified in the description associated with the term.

Let us consider what follows about reference change if meaning is governed by semantical rules in this manner. In the quote before last (1981e, p. 98), Feyerabend gives a disjunctive specification of meaning change. The first way for meaning to change is "if a new theory entails that all concepts of the preceding theory have zero extension". Since on the present account concepts are determined by rules, this means that the rules of the older theory appear to be unsatisfied from the vantage-point of the later theory. More exactly, it implies that the rules of the two theories are jointly incompatible. The second way in which meaning may change is if the new theory "introduces rules which cannot be interpreted as attributing specific properties to objects within already existing classes, but which change the system of classes itself". On this second alternative the rules of the new theory are inapplicable to members of the old classes; instead, they determine a whole new classificatory system.

In either case the transition between theories involves discontinuity of reference. For if reference is determined by rules, then in order for there to be common reference the same objects must satisfy different systems of rules. But on the first alternative the same objects cannot satisfy both sets of rules: incompatible sets of rules are not jointly satisfiable. While on the second alternative no common objects can belong to both systems of classes. For if the new rules attribute no properties to members of old classes, then no criterial property specified in a new rule can be instantiated by any members of an old class. The systems of classes must be completely disjoint. Thus, in both cases there can be no common reference between incommensurable theories, so the transition between such theories is referentially discontinuous.

In sum, Feyerabend is committed to discontinuity of reference between incommensurable theories. This is due to his assumption that the mutual exclusiveness of the concepts of incommensurable theories is relevant to reference. More particularly, it is because he assumes such conceptual incompatibility to preclude the possibility of co-reference.

The problem, as noted earlier, is that Feyerabend is also committed, by the pragmatic view of observation, to continuity of reference to observable objects. For the pragmatic account allows preservation of the use of observational vocabulary in the transition between theories. Thus the reference of observational terms may remain invariant between incommensurable theories, which is inconsistent with there being discontinuity of reference between such theories.

Let us see whether this tension can be resolved. If both the pragmatic theory of observation and the incommensurability thesis are retained, there appear to be just two options available. The first option is to minimize the discontinuity of reference attendant upon incommensurability in order to remove the conflict with the pragmatic account. The second option is to deny that stability of the pragmatic features of the use of observational language is relevant to the issue of reference.

The first option confines discontinuity of reference to the theoretical level. That is, it might be granted that observational terms may preserve reference through change of theory. In that case, failure of interdefinability of theoretical concepts would at most lead to co-reference failure between theoretical expressions. This would remove the tension by allowing continuity of reference at the observational level in conjunction with discontinuity of theoretical term reference.

There are two problems which prevent this approach from being incorporated into Feyerabend's position. In the first place, according to Feyerabend, it is not just the meaning of theoretical terminology which is determined by the basic principles of a theory. Rather, in his view observational language also receives meaning from the theory in which it is employed. Thus the argument that terms which are defined on the basis of incompatible principles are not interdefinable applies to observational vocabulary as well as to theoretical terms. So if the basic principles of a pair of theories are incompatible, then observational terms which belong to such theories cannot be brought into analytical relations with each other. Thus by the argument that incompatibly defined terms fail to co-refer, it follows that the observational terms of incommensurable theories do not co-refer.

In the second place, the first option introduces an asymmetry between the way reference is determined for observational and theoretical terms. For according to this approach, the reference of observational terms is independent of theory, while that of theoretical terms is determined by theory. This assumes that the reference of theoretical terms is determined by description while that of observational terms is determined by direct attachment. But nothing in Feyerabend warrants this, since his treatment of the issue presupposes a description theory of reference.

The second option is to take pragmatic factors to be irrelevant to reference. This is to deny that pragmatic stability entails continuity of observational reference. This removes the tension by denying the conflict between invariance of observational use and discontinuity of reference. It does so by asserting that there may be continuity of the pragmatic features of observational use even though there is complete discontinuity of reference.

To put the point in Feyerabend's terms, this is to treat reference as a semantic rather than a pragmatic issue. Feyerabend characterizes observation sentences pragmatically in terms of the circumstances of their application, as opposed to semantically in terms of their meaning. The present idea is that reference is semantical and is not affected by pragmatic factors.

To say that reference is semantical is to say that a term's reference is determined by its meaning. Thus, in view of the relations between their concepts, incommensurable theories are fully disjoint with respect to reference. Since observational meaning depends on theory, not even the extensions of their observational terms overlap.

The distinction between pragmatic and semantical aspects of language use enables it to be said that theories which do not refer to any of the same entities may still be about the same things. They may be applied to the same domain in a pragmatic sense even if in a strict semantical sense they fail to have any common reference. For it is a pragmatic fact about the empirical setting in which such theories are employed that they are brought to bear on the same phenomena.

It is also a pragmatic fact that the conditions in which an observational vocabulary is employed remain stable through change of theory. But on the present approach, such pragmatic stability is not taken to affect the semantical issue of whether observational vocabulary continues to have the same reference. For reference depends on meaning, not the conditions in which a term is employed.

From a pragmatic point of view, an observation sentence may exhibit the same features regardless of the theoretical context in which it occurs. But, on the present approach, the meaning and reference of observational vocabulary is sensitive to theoretical context. Opposing theories applied to the same empirical domain may be associated with the same observational vocabulary and may employ the same observational sentences. Yet the ontologies of such theories may differ massively; the entities posited by one may be rejected by another. Thus, from a semantical viewpoint, the vocabulary under the interpretation of one theory will seem to fail to refer when considered from the perspective of the other. Because of the conflict between the presuppositions of the meanings of such terms, it appears that they do not denote the same entities even though they are pragmatically applied to the same domain.

The distinction between reference as a strict semantic notion and the pragmatic idea of being applied to the same domain removes the tension between incommensurability and the pragmatic account of observation. For on the strict semantical analysis of reference, pragmatic stability affords no basis for the semantic continuity of reference. The pragmatic conditions of linguistic use at the observational level may survive changes of meaning which cause reference to change. So the transition between incommensurable theories can be a referentially discontinuous one in spite of stable pragmatic conditions of observational use.

Unlike the first option, this second approach does cohere with Feyerabend's position. It is consistent with his view of reference as determined by rules or descriptions. And it is consistent with his view that observation statements may be characterized pragmatically. However, the second option is objectionable in its own right.

First of all, it overlooks the role of pragmatic determinants of reference. In order to deny continuity of reference for observational terms, it must be denied that reference can be fixed directly by means of pragmatic relationships such as ostension. In effect, to sustain the idea that reference may be discontinuous at the observational level, comprehensive use must be made of a description theory of reference. That is, it must be assumed that the reference of an observational term is determined solely by the description which specifies its meaning.

For suppose that an observation language is employed by a pair of incommensurable theories. Because the meaning of observational vocabulary is assumed to depend on the theory in which it is employed, it would follow that the definition of any observational term defined in one theory would be incompatible with fundamental principles of the other theory. To deny that any observational term of one theory may co-refer with observational terms of the other theory, such incompatibility must be taken to imply failure of co-reference. But that assumes that the reference of observational terms is determined by the descriptions which define them.

But this is problematic, for it leads to a denial of ostensive determination of reference. Consider an observational term occurring in a pair of incommensurable theories which has a unique ostensive definition. It follows from the above that the extension of the tokens of the term occurring in one theory can have no members in common with the extension of tokens of the term which occur in the other theory. Even if the ostensive definition of the term is the same in both theories, the term as used in either theory has completely distinct reference. But to deny that tokens of a term ostensively defined in the same way may co-refer is to deny that the reference of an observational term may be determined by ostension. And to deny that ostension may determine reference is to reject the fundamental means by which linguistic expressions are linked referentially to the extralinguistic world.

A second problem with emphasis on semantical at the expense of pragmatic factors is the apparent consequence that there can be no conflict between the content of incommensurable theories. The problem is that conceptually disparate theories do appear capable of such conflict; and this depends at least in part on their pragmatic application to the same domain.

Let us consider the case of 'impetus' again. As we saw earlier, Feyerabend rejects the attempt to correlate the concept of impetus with Newtonian concepts by means of a hypothesis of co-extensiveness such as 'impetus=momentum'. If no pragmatic factors are relevant to reference, it would appear to follow that there is no relation of co-reference linking either observational or theoretical expressions of the two theories. That is, it would seem that the content of such theories has nothing whatsoever to do with each other. In particular, it would appear to follow that no sentence about impetus may conflict with any sentence about momentum. For in order for the truth of a sentence about the impetus of an object to be incompatible with that of a sentence about the object's momentum, the terms 'impetus' and 'momentum' must at least have overlapping reference.

The problem with taking such theories to be completely unrelated to each other is that sentences about impetus and momentum may be applied to the same phenomena. As a pragmatic fact about the circumstances in which they are applied, the terms 'impetus' and 'momentum' may be predicated of the very same bodily motion. According to the above reasoning, such predications do not conflict with each other because the terms do not have the same extensions. But it is unnecessary to reject

Feyerabend's view that the two terms cannot co-refer in order to show that sentences applying the two terms to the same object are incompatible.

The denial that the terms 'impetus' and 'momentum' co-refer is not simply the claim that their extensions are disjoint. It is rather a denial that the terms could refer to the same thing. As a pragmatic fact, both terms are applied to the same objects, viz. projectile bodies. Because they neither co-refer nor refer to separate things, such joint application implies that at least one fails to refer. Now, 'impetus' cannot refer from a Newtonian perspective because the existence of impetus is incompatible with Newtonian assumptions about projectile motion. But this implies that the truth of a sentence which predicates 'impetus' of an object is incompatible with Newton's physics. Taken together with the pragmatic fact of joint application, the incompatibility of the concept of impetus with Newton's laws therefore entails the existence of a conflict between the theories.

5.3 Kuhn's original position

Kuhn's views have changed and several stages in the development of his position on reference change may be distinguished. This section examines Kuhn's original position in his (1970a).

The development of the position may be brought out by briefly tracing the changes in Kuhn's handling of the metaphor of "world-change". Originally, he suggested that in some sense "when paradigms change, the world itself changes with them" (1970a, p. 111). This depicted paradigm change as a form of space travel. "It is rather as if", he said, "the professional community had been suddenly transported to another planet" (p. 111). But he soon changed his tone and it emerged that, in so speaking, his point had been to stress alterations in basic ontological categories (1970b, pp. 269-70, 275). So he came to moderate the metaphor, choosing to say instead that "languages cut up the world in different ways" (1970b, p. 268). It was only later, when the role of the metaphor was largely taken over by other formulations, that the position came into focus. Now little remains of the original image except the claim that "languages impose different structures on the world" (1983, p. 682). It has largely been replaced by new turns of phrase, such as "homology of lexical structure" and "taxonomic structure" (1983).

If it is not a metaphor, the world-change image loses its point: literally it defies belief, but figuratively it is a forceful image.[7] The point of the image is that paradigms differ so radically that it is as if they were about different worlds. Since scientists do not literally change worlds when they convert to a new paradigm, the question arises of how to interpret the notion of reference in the context of the world-change metaphor.

Let us consider the following passage, which is Kuhn's most explicit original treatment of reference:

> the physical referents of these Einsteinian concepts [i.e. space, time and mass] are by no means identical with those of the Newtonian concepts that bear the same name. (Newtonian mass is conserved; Einsteinian is convertible with energy. Only at low relative velocities may the two be measured in the same way, and even then they must not be conceived to be the same.) (1970a, p. 102)

He adds that Newton's laws are not limiting cases of Einstein's because in the transition between them "we have had to alter the fundamental structural elements of which the universe to which they apply is composed" (p. 102).

To say that both Einsteinian and Newtonian concepts refer and that in the transition between them "structural elements" of the universe have had to be changed suggests the existence of referents to which the concepts of the rival paradigms successfully refer. That is, it seems to imply that paradigms possess their own referents and that the change between paradigms involves a change in the nature of the world to which the paradigms refer. But the passage is ambiguous and the reading which best fits Kuhn's later discussion is neither the only nor the best interpretation available. Consequently, no unequivocal analysis of Kuhn's original view of radical reference change can be given.

The following procedure, not designed to yield a definitive interpretation, will be adopted. It will be assumed that Kuhn's position is ambiguous and that there is more than one acceptable interpretation of it. Three distinct interpretations will be given. The first reading to be given is the one that fits best with Kuhn's later position. But it will be shown that a second interpretation is to be preferred to the first one. Criticism of this second position will be given, which will lead to the formulation of a third interpretation. It will then be shown that, understood in this third way, Kuhn's point depends on a description theory of reference and leads to unacceptable consequences.

On the face of it, the first of the last two quoted passages has the following two implications about Einsteinian and Newtonian terms for mass. First, it implies that both Newtonian and Einsteinian 'mass' do in fact succeed in referring to their respective kinds of mass. Second, it implies that the two terms do not have the same reference.

The first interpretation is consistent with Kuhn's later position, which will be discussed in 5.4. In his later position, Kuhn associates classificational change with reference change: general terms denoting classes undergo extensional change when the system of classificational groupings is altered. With this later view in mind, it is possible to interpret Kuhn's treatment of the two concepts of mass in the above passage as follows. The Einsteinian and Newtonian concepts of mass may be taken to belong to distinct systems of classification of the objects and processes studied by physics. As such, the two concepts are different means of classifying a shared domain of physical objects. Both terms successfully refer, yet because they classify differently they do not refer to exactly the same class of objects.

This way of reading the passage is suitable on three counts. First, the expressions 'Newtonian mass' and 'Einsteinian mass' suggest that there is a set of objects which constitutes the extension of the Einsteinian and Newtonian terms respectively. Second, Kuhn says that those referents are "by no means identical", which could be taken to mean that they are not exactly the same class of objects. Third, the comment that both kinds of mass may be measured in the same way at low velocity suggests that the extensions of the two terms intersect with respect to some subset.

Thus the passage could be taken to mean that Einsteinian and Newtonian 'mass' successfully refer to their own respective reference classes which partially overlap with each other.[8] Kuhn's later position could therefore be superimposed upon the above passage in order to provide retrospective clarification of his original view.

This interpretation of the passage is, however, problematic. It depends on taking the phrase "by no means identical" in a weak and slightly unnatural sense which allows there to be an overlap of reference. The phrase is consistent with a stronger interpretation on which the extensions have no common members at all. For to say that two things are by no means identical tends more clearly to suggest that absolutely nothing is shared by them. Moreover, this stronger sense is equally consistent with the passage. For it is clear from the first parenthetic remark that the two sorts of mass are considered to be entirely different in nature: one is conserved while the other converts with energy. So there is at least as much reason to take it that the two terms have quite distinct extensions which do not intersect even partially.

Yet locutions such as 'the physical referents of Einsteinian [or Newtonian] concepts' or 'Newtonian [or Einsteinian] mass' seem to imply that both of the terms do have reference. Given the stronger interpretation of "by no means identical", this would in turn imply that Einsteinian and Newtonian 'mass' do not co-refer at all even though both refer. This cannot be the case, for it would then follow that at least one of the two terms fails to refer to any ordinary physical objects which possess mass. That is, it implies that either Einsteinian or Newtonian 'mass' fails to denote everyday masses and yet succeeds in referring. But neither term can refer successfully if it in fact fails to refer to stereotypical masses.

In sum, the first interpretation is strained because it allows extensional overlap. The phrase "by no means identical" seems to rule out common reference altogether. But then it is unclear how to combine non-overlap with successful reference. The second interpretation, to which I now turn, attempts to combine co-reference failure with successful reference by making use of Kuhn's metaphor of "world-change".

To say that Einsteinian and Newtonian 'mass' cannot both refer and refer disjointly amounts to saying that they cannot both refer to objects in the same world. But what if they refer to objects in different "worlds"? Given Kuhn's use of the "world-change" metaphor, the possibility that paradigms are about their own "worlds" demands consideration. If the metaphor is taken as a metaphor, it might be said that paradigms are associated with "worlds". Such "worlds" are determined by the ontology of the paradigm and reference, for each paradigm, is relative to "world".

This second interpretation accords with what seems to be a popular way of reading Kuhn. It appears to be widely held that Kuhn's position is that there is a radical discontinuity of reference between incommensurable paradigms, because the terms of each paradigm refer to objects within its own "world".[9] Such an interpretation has the merit of consistency with the passage being examined. For if it can be satisfactorily explicated, it would enable the Einsteinian and Newtonian uses of 'mass' to refer successfully without co-referring.

But what is it for the terms of a paradigm to refer to a "world"? It may be assumed for present purposes that Kuhn is a realist for whom the world independent of theory does not itself change in the transition between theories.[10] So reference in a "world" will not be the same thing as reference to actual objects in the world.

Instead of thinking of the "world" of a paradigm as the world itself, we might think of it strictly in terms of the paradigm's ontological commitments. That is, the "world" of a paradigm is the set of objects which exist according to the paradigm. If we take this to mean that the "world" of a paradigm is the way the world would be if the paradigm were true, then it is plausible to approach the idea in the following manner. In a discussion of Kuhn's paradigms, Cedarbaum (1983) interprets the concept of a paradigm using the notion of a semantical model. Though Cedarbaum does not explicitly equate the "world" of a paradigm with its model, it is a natural extension of his idea to analyze the notion of a "world" in terms of that of a model.

Cedarbaum suggests that the concept of a paradigm should be understood as follows:

> The essential constituents of a paradigm, for Kuhn, are an axiom system and a model (in the technical sense) of that system. (1983, p. 204)

According to Cedarbaum, the technical sense of 'model' is defined this way:

> A 'model' of a theory is a logical interpretation of the theory, a choice of a universe of variables over which the quantifiers range and an assignment of denotations to the term letters, under which all of the axioms of the theory are true. (1983, p. 210)[11]

To illustrate, Cedarbaum compares the semantical features of 'sun' and 'planet' with regard to the models of Ptolemaic and Copernican astronomy:

> The term 'sun', for instance, on the traditional view of naming, has a different intension in the Copernican model than in that of Ptolemy, and the word 'planet' has both a different intension and a different extension in one model than in the other. (1983, p. 205)

Presumably, the extension of 'planet' includes the earth in the Copernican model but does not include it in the Ptolemaic. So, if reference is understood as relative to a model, the shift between the two theories involves a change in the reference of 'planet'.

If we equate the meaning of the term 'world' as used by Kuhn with that of 'model' in this sense, then reference within the "world" of a paradigm becomes reference within its model. Then we might follow Cedarbaum when he claims that, according to Kuhn:

> The model for the Newtonian theory is not a submodel of the model for the Einsteinian theory. In the former, for example, 'mass' refers to a substance which is conserved, while in the latter it refers to a substance which is convertible with energy. (1983, p. 206)

This approach accords with the present interpretation of Kuhn's view of reference in terms of reference relative to the "world" of a paradigm. Instead of talking of reference relative to a paradigm we may speak of reference in a model. On this approach, incommensurable theories have disjoint models. Because reference is understood as restricted to the model of a theory, the terms of incommensurable theories consequently do not co-refer.

Analysis in terms of models is more perspicuous than talk of "worlds". Moreover, it reconstructs a significant part of what Kuhn meant to convey with the "world-change" metaphor. The "world" of a paradigm is the way the world is according to the paradigm's ontology. It is the way the world would be if the paradigm's ontological commitments actually obtained. In this respect the notion of a model captures what is essential to the notion of a "world".

But there is a serious shortcoming with the notion of a model which prevents it from being of any real use in the present context: namely, that nothing follows from the notion of a model about extralinguistic reference. If a theory has a model its terms do not necessarily refer; a model is how things would be if they did refer. If the notion of a "world" is construed in terms of the notion of a model, then the question of extralinguistic reference remains completely unanswered.

Models reflect theory rather than world and do not necessarily correspond to reality. The model of a theory is determined by the semantical truth conditions of a theory; it is a semantic interpretation of the theory, relative to some (actual or non-actual) domain, on which the theory is true. If the theory is false or if its terms do not refer, then the actual world does not constitute a model of the theory. Thus the question of whether a model corresponds to reality reproduces the question of whether the theory's terms actually refer. To say that a term refers in some model is not yet to say that it really does refer, for it remains to ask whether the actual world is a model of the theory. So the fact that Einsteinian or Newtonian 'mass' refer within the models of their respective theories does not decide the question of whether they refer outside of a model to actual masses.

Thus the second interpretation of the Kuhn passage does not yield the required thesis about reference. What is needed is an account on which both terms really do refer. Talk of models and metaphorical "worlds" does not provide such an account. Nor does such talk provide an account of radical reference change. For if the question of reference remains unanswered, then so does that of reference change.

The only way this interpretation can have the consequence either that both terms refer or that there is discontinuity of reference would be to combine it with a rejection of extralinguistic reference. If there is no model-transcending reference to extralinguistic entities, then it might be said that the terms employed by paradigms refer to objects in the model of their paradigm. Then paradigms would be referentially disjoint and the transition between them referentially discontinuous. As against this approach, however, it will be argued in section 7.6 that such an outright rejection of reference is problematic in its own right.

I will now propose an interpretation according to which Kuhn is not committed to the actual reference of both Einsteinian and Newtonian 'mass'. On this third interpretation, Kuhn employs a description theory of reference on which difference in the definition of the two terms is relevant to their reference. According to the analysis of the two terms which is entailed by this interpretation, Newtonian and Einsteinian occurrences of 'mass' do not co-refer and at least one of them fails to have reference.

The novelty of this third interpretation consists in its analysis of Kuhn's use of the term 'reference'. The proposal is that Kuhn's word 'referent' not be taken literally. It is not to be read as 'referent' and is not to be taken to imply reference at all. Instead, it may be glossed as 'purported referent' or 'putative referent', which makes the question of actual reference a separate issue. Similarly, the locutions 'Einsteinian mass' and

'Newtonian mass' are not to be understood to entail the existence of different sorts of mass, but a difference in purported reference. In parallel fashion, Kuhn's talk of "worlds" may be replaced by a locution such as 'the way the world is purported to be'.

So, in particular, let us take the term 'referent' in the Kuhn passage above as elliptical for 'putative referent'. Thus Kuhn's claim that the "physical referents" of Einsteinian and Newtonian 'mass' are "by no means identical" may be read as a statement about what they putatively refer to. Namely, the terms purport to refer to quite distinct things. More precisely, the descriptions which specify what purport to be the referents of Einsteinian and Newtonian 'mass' are mutually inconsistent.

Kuhn's treatment of reference in the passage under analysis suggests tacit acceptance of a description theory. His parenthetic argument against the co-reference of the tokens of 'mass' appeals to disagreement over the conservation of mass. For, as evidence that they do not co-refer, he notes that "Newtonian mass is conserved; Einsteinian is convertible with energy". This assumes that the inconsistency of these two descriptions is evidence against co-reference, which assumes descriptive content is relevant to reference. Moreover, his final parenthetic remark that the two masses can be measured in the same way at low velocity but "must not be conceived to be the same" is also given as evidence against co-reference. Not conceiving of something as the same as some other thing is only relevant if the way something is conceived of is relevant to reference. To assume that the latter is of relevance to reference is to assume that descriptive content determines reference.

The present interpretation combines this description-theoretic analysis of Kuhn's view of reference with the reading of Kuhn's word 'referent' as 'putative referent'. The interpretation squares with Kuhn's view that Einsteinian and Newtonian occurrences of 'mass' do not have the same extension. For on the assumption of a description theory of reference, the inconsistency of the descriptions precludes co-reference. Moreover, the interpretation of 'referent' as 'putative referent' removes the implication, conveyed by Kuhn's use of the word 'referent', that both of the terms do in fact refer. Thus it removes Kuhn's apparent commitment to the concurrent existence of Newtonian and Einsteinian masses.

Such an interpretation in terms of a description theory can, however, be shown to have consequences which are themselves unacceptable. The fundamental difficulty is that it leads to an improper analysis of the relation between the Einsteinian and

Newtonian concepts of mass. For it implies that the two concepts are unable to function as conflicting characterizations of a common physical property, but instead constitute concepts of quite distinct quantities.

Let us consider what follows if the reference of Einsteinian and Newtonian 'mass' is determined by description. Given Kuhn's remarks about the difference between the concepts, this would presumably imply that the Einsteinian term 'mass' (putatively) refers to a magnitude which is convertible with energy while Newtonian 'mass' (putatively) refers to a magnitude which is conserved. Since the same property cannot both be convertible with energy and not be so convertible, it follows that the terms do not refer to the same property. Thus, the assumption that reference is determined by description in this way has the consequence that the Einsteinian and Newtonian concepts cannot be opposing characterizations of the nature of a single physical magnitude referred to by both theories as 'mass'.

This consequence has a number of objectionable features. In the first place, it implies that Newton's theory cannot be contradicted by denying that mass is a conserved quantity. Given that the Newtonian concept of mass is a concept of mass which is conserved, to deny that mass is conserved is ipso facto not to speak about the same kind of mass. In the second place, if Newtonian 'mass' only refers to mass if it is conserved, then certain sorts of experimental results are ruled out altogether. For example, it would be impossible to discover empirically that the mass referred to in Newtonian explanations of physical motion is convertible with energy. In the third place, given that the two terms fail to co-refer for conceptual reasons, it would be impossible for both theories to be partially correct accounts of mass. For reasons such as these, application of the description theory appears to yield a mistaken analysis of the relation between the two concepts of mass.

There is no need, however, to impose a description-theoretic analysis on the relation between the two concepts. For it may be denied that the reference of 'mass' is determined by theoretical description. Instead, we may say that both theories are about mass and that they predicate different properties of the physical magnitude which both refer to as 'mass'. Thus the two theories disagree about the nature of mass and their opposing descriptions of mass constitute more or less correct accounts of it. On such a view, there is no radical discontinuity of reference in the transition between the two concepts of mass, for the term 'mass' retains its reference. There is a major change of theory and an associated evolution of the concept of mass; but the two theories remain conflicting accounts of the same thing.[12]

5.4 Kuhn's later view

Kuhn's later treatment of reference change resembles the first interpretation of his original position discussed in 5.3. It is a restricted thesis according to which the reference of key categorial terms varies in the context of overall referential stability.

Briefly, Kuhn's later position associates change of reference with classificational change. Languages and theories incorporate systems of classification which organize and classify objects into categories. Systems of classes or categories may differ as to how they divide objects into classes. The language associated with a classificatory system may be altered semantically if the system is transformed. Terms with a given extension within one classificational system may acquire a new extension in being moved to another system.

In Kuhn's later writing, there is a significant recasting of the "world-change" metaphor. In its later use the image may be interpreted in terms of categorial change: the term 'world' refers to a system of categories, and 'the world of a theory' is a theory's categorial system. Kuhn says, for example, that "languages cut up the world in different ways" (1970b, p. 268) and that "languages impose different structures on the world" (1983, p. 682). Presumably, the "world" of a theory is the system of categories or classes produced by cutting up or structuring the world. At the level of metaphysics, the recast metaphor is indicative of a position weaker than the one originally suggested by talk of "world-change". For to say that the world can be cut up in various ways implies that there is one world and that it is divisible in a variety of ways.

Kuhn's concern with the difference between the systems of categories employed by languages surfaces explicitly in his (1970b) with his comment that "languages cut up the world in different ways" (p. 268). He elaborates the point with reference to the example of 'gavagai', Quine's imagined native word for rabbit:[13]

> Quine points out that, though the linguist engaged in radical translation can readily discover that his native informant utters 'Gavagai' because he has seen a rabbit, it is more difficult to discover how 'Gavagai' should be translated. Should the linguist render it as 'rabbit', 'rabbit-kind', 'rabbit-part', 'rabbit-occurrence', or by some other phrase he may not even have thought to formulate? (1970b, p. 268)

In using Quine's example to illustrate the idea of alternative linguistic orderings of the world, Kuhn's point is that the concepts under which objects are classified depend on conceptual scheme. For instance, a native language with a fundamental ontology of occurrences or parts of things may conceive rabbits as occurrences of rabbithood or as collocations of rabbit-parts.

There is a formal similarity between the gavagai example and the case of mass discussed in 5.3. In both cases, alternative conceptual frameworks are brought to bear on the same set of objects, viz. rabbits or material bodies. In both cases the set of objects is invariant, though the concept under which the object is subsumed varies with framework.

Most of the examples discussed by Kuhn, however, exhibit a different kind of relationship. In general, Kuhn discusses cases in which one or more object or sets of objects is transferred from one natural category to another, thus altering the membership of the categories. For example, Kuhn tells us that:

> Dalton's atomic theory ... implied a new view of chemical combination with the result that the line separating the referents of the terms 'mixture' and 'compound' shifted; alloys were compounds before Dalton, mixtures after... Whatever the reference of 'compound' may be, in this example it changes. (1970b, p. 269)

In this passage Kuhn claims that the reference of a pair of general terms changes due to the transfer of a subset between their extensions.[14] Such a change of extension constitutes a change of higher-order categories in the context of lower-order order categorial invariance. The example is partially analogous with the gavagai and mass cases at the level of the transferred set: the category of the alloys remains stable throughout the change.

The change of reference discussed in the above passage is limited because it combines change of classification with preservation of reference. The restricted nature of the reference change thesis which this implies is further exemplified by the following passage, also from Kuhn's (1970b):

> normal science depends [on] a learned ability to group objects and situations into similarity classes... One aspect of every revolution is, then, that some of the similarity relations change. Objects which were grouped in the same set before are grouped in different sets afterwards and vice versa. Think of the sun, moon, Mars, and earth before and after Copernicus; of free fall, pendular and planetary motion before and after Galileo; or of salts, alloys, and a sulphur-iron filing mix before and after Dalton. Since most objects within even the altered sets continue to be grouped together, the names of the sets are generally preserved. Nevertheless, the transfer of a subset can crucially affect the network of interrelations among sets. Transferring the metals from the set of compounds to the set of elements was part of a new theory of combustion, of acidity, and of the difference between physical and chemical combination. (1970b, p. 275)

As with the preceding quotation, the reference change described in this passage is restricted in scope. Not all terms are affected by change of reference, and those which are affected may partially preserve their extensions. A set of objects which is transferred from one set to another may be referred to under both systems of classification. The terms which refer to the transferred objects and sets may themselves preserve their reference through change of classification.

To judge from Kuhn's examples, the alleged changes of reference appear to be confined to higher-order categories. For while the putative reference of the general term 'planet' changes, the singular term 'earth' continues to denote the earth. In the Dalton case, the extension of 'alloy' is constant while the classification of the alloys as compound or mixture varies. In such cases, the change of reference occurs at the level of the more general categorial terms, such as 'compound' or 'planet', to whose extensions the alloys or the earth belong.

Kuhn has a more explicit discussion of continuity of reference elsewhere. In his (1979) he endorses the causal theory of reference as applied to individual objects. But he raises objections to its account of the reference of kind terms. He adopts the position that reference to individual objects may be preserved while reference to kinds is subject to change.

Kuhn's objection to the causal theory concerns the role of ostension in fixing reference. He grants that, "in the case of proper names, a single act of ostension suffices to fix reference" (1979, p. 412). But he argues that the reference of a kind term cannot be fixed simply by an act of ostension:

> if I were to exhibit to you the deflected needle of a galvanometer, telling you that the cause of the deflection was called "electric charge," you would need more than good memory to apply the term correctly in a thunderstorm or to the cause of the heating of your electric blanket. Where natural-kind terms are at issue, a number of acts of ostension are required. For terms like "electric charge," the role of multiple ostensions is difficult to make out, for laws and theories also enter into the establishment of reference... [E]stablishing the referent of a natural-kind term requires exposure not only to varied members of that kind but also to members of others — to individuals, that is, to which the term might otherwise have been mistakenly applied. (1979, pp. 412-3)

In so arguing, his objection appears to be twofold. First, reference to a kind cannot be fixed by ostension of a single object, for such ostension cannot itself determine what does and what does not belong to the kind. Second, in fixing the reference of terms for unobservable kinds pure ostension must be supplemented by theoretical description.

There is no need to criticize Kuhn's rather sketchy remarks on the causal theory of reference in detail. But the following points should at least be briefly noted. As we saw in Chapter Two, the causal theory of reference can allow a role to multiple ostensions in establishing reference. It may also grant a role to descriptions in fixing the reference of theoretical and kind terms. More importantly, Kuhn appears to confuse the issue of what is required to learn the use of a term with what is required to fix its reference. It is true that learning the expression 'electric charge' as applied to a galvanometer will not enable one to apply the expression in situations for which one has received no instruction. Nevertheless, the extension of 'electric charge' may still be established by fixing it as the physical magnitude measured by a galvanometer. These reservations notwithstanding, Kuhn's objections are in accord with the view that the causal theory needs more than a simple ostensive model of reference-fixing (see sections 2.5-2.6).

In the following passage, Kuhn puts forward the thesis that reference to individual objects may be preserved while their classification changes:

> the causal theory of reference [is] a significant technique for tracing the continuities between successive theories and, simultaneously, for revealing the nature of the differences between them... The techniques of dubbing and of tracing lifelines permit astronomical individuals — say, the earth and moon, Mars and Venus — to be traced through episodes of theory change, in this case the one due to Copernicus. The lifelines of these four individuals were continuous during the passage from heliocentric to geocentric theory, but the four were differently distributed among natural families as a result of that change. The moon belonged to the family of planets before Copernicus, not afterwards; the earth to the family of planets afterwards, but not before. (1979, pp. 416-7)

The nature of the change of classification that takes place is more fully described in a later paper (1981):

> Before [the shift from Ptolemaic to Copernican astronomy] occurred, the sun and moon were planets, the earth was not. After it, the earth was a planet, like Mars and Jupiter; the sun was a star; and the moon was a new sort of body, a satellite. (1981, p. 2)

On the basis of these two quotations, Kuhn does appear to hold a differential reference change thesis: the extensions of kind terms vary, but reference to the individual items in their extensions is preserved.

There are a number of problems with interpreting Kuhn's thesis of the preservation of reference of names. The first is due to a peculiarity of the astronomical example: viz. the sun, moon and planets are all observables. Does this suggest that the thesis applies only to the names of observable objects? If so, then it would appear to follow that singular reference to unobservables varies with theory.

However, there is no need to interpret Kuhn's thesis as applying exclusively to names for observable objects. For Kuhn's objections to the causal theory concern reference to kinds; they raise no difficulties about reference to particular entities, observable or otherwise. Moreover, he does allow that reference to an unobserved cause may be secured by appeal to its observed effects. For while he denies that the full extension of 'electric charge' can be fixed by the description of the cause of a single ostended effect, he grants that singular reference to the unobserved cause of the particular effect may be secured:

> pointing to a galvanometer needle while supplying the name of the cause of its deflection attaches the name only to the cause of that particular deflection (or perhaps to an unspecified subset of galvanometer deflections). It supplies no information at all about the many other sorts of events to which the name 'electric charge' also unambiguously refers. (1979, p. 411)

Given that singular reference to a particular unobservable cause is acceptable to Kuhn, he would appear to have no basis on which to deny singular reference to unobservables.

A further question of interpretation arises because of the emphasis in Kuhn's (1979) on stability of reference to individuals rather than kinds. The astronomical example involves stability of reference to individual bodies and change of extension at the level of the class of planets. But, in contrast, Kuhn's Dalton example concerns the transfer of the set of alloys from the compounds to the mixtures: the kind term 'alloy' refers to the set of alloys throughout the re-classification. Given Kuhn's view that categories change with change of theory, the question arises of why that set should not be altered.

Part of the answer to this appears to be rather simple. Namely, Kuhn's thesis is a limited rather than a radical reference change thesis. So it suffices to say on Kuhn's behalf that change of theory need not alter every class of the classificational system.

But this does raise the question of how stability of reference to kinds differs from stability of reference to individuals. For Kuhn, apparently, what it takes for the extension of a kind term to remain stable differs from what it takes for reference to an individual to be stable. What makes the extension of 'alloy' remain the same differs from what makes the reference of 'earth' stable. The former is due to the retention of a belief in the homogeneity of the class of alloys, while the latter is due to continued application of the term to the same object to which it was originally attached.

Why is the reference of kind terms determined differently from that of names? As we saw earlier, Kuhn denies that the reference of a kind term may be fixed by simple ostension in the way that a name's referent may be fixed. Thus, the causal theory ensures stability of reference to individuals, not to kinds. This is because of the difference between individuals and kinds:

> When one makes the transition from proper names to the names of natural kinds, one loses access to the career line or lifeline which, in the case of proper names, enables one to check the correctness of different applications of the same term. The individuals which constitute natural families do have lifelines, but the natural family itself does not. (1979, p. 411)

Kuhn's point is that with a name one can check to see if an individual is the same as the one to which the name was attached in an original naming ceremony. But kinds contain discrete members, and ostension does not itself determine which objects, other than those present at the naming ceremony, constitute the extension of the kind.

Kuhn's point appears to be an epistemic one.[15] Namely, to find out which objects belong to a kind, aside from those in the ostended sample-set, must involve a theory about the membership of the kind. This is also suggested by Kuhn's view that the extensions of categorial terms such as 'planet' are subject to variation with change of theory. For the extension of such terms changes because the theory of what belongs to such categories changes. In discussing the re-classification of astronomical bodies by Copernican theory, Kuhn says:

> That sort of redistribution of individuals among natural families or kinds, with its consequent alteration of the features salient to reference, is, I now feel, a central (perhaps the central) feature of the episodes I have previously labeled scientific revolutions. (1979, p. 417)

Moreover, in his (1981) he suggests that:

> What characterizes revolutions is, thus, change in several of the taxonomic categories prerequisite to scientific descriptions and generalizations. That change, furthermore, is an adjustment not only of criteria relevant to categorization, but also of the way in which given objects and situations are distributed among pre-existing categories. (1981, p. 25)

Kuhn's view appears to be that the extension of kind terms depends on theory. Whereas reference to an individual object may be determined by an original naming ceremony, reference to kinds depends on classification and changes with classificatory change.

Now this does appear to make reference overly dependent upon classification. For it seems not to allow the possibility that the alloys were wrongly classified as compounds and that the extension of 'compound' did not in fact change.[16] And it appears to underestimate the extent to which the reference of theoretical terms may be independent of particular theories. I will return to these points later in the section.

Let us adopt the following interpretation of Kuhn's view: he holds that the reference of kind terms depends on theory because theory is needed to group objects into kinds. On this interpretation of his position, however, Kuhn can be shown to be committed to stability of reference for low-level kind terms.[17] This is because establishing stability of singular reference to objects must also introduce theoretical considerations.

In a passage which was quoted earlier, we find Kuhn stating that: "the techniques of dubbing and of tracing lifelines permit astronomical individuals — say, the earth and moon, Mars and Venus — to be traced through episodes of theory change" (1979, p. 417). The example of the heavenly bodies is in fact a case in which considerable theoretical apparatus must be brought into play. For in order to "trace the lifeline" of an "astronomical individual", an astronomical theory is required.

Kuhn appears to accept, for example, that the term 'Mars' still refers to the same object that it referred to in earlier use originating from the name's introduction. But to determine that Mars is the original referent, it must be established which object 'Mars' originally denoted. And this in turn requires a theory which provides information about the past movement and position of the planet Mars. It is also necessary to show that present use continues to apply the term to the same body. Such determination of past and present reference for 'Mars' requires a theory about the behaviour of heavenly bodies.

To identify an observed phenomenon as the planet Mars requires rudimentary knowledge of astronomy. By contrast, to determine that distinct objects — dogs, say, or trees, or pieces of metal — are members of a single kind need not involve extensive theory. Since Kuhn is prepared to accept the causal theory of reference as an account of the stability of reference to Mars, he should therefore accept it as well for less theory-laden kind terms. By his own concession to the causal theory, Kuhn appears committed to stability of reference, by and large, for low-level empirical kind terms.

I will now discuss a complication with Kuhn's position which arises because of Kuhn's emphasis on names as opposed to descriptions. Singular reference occurs via one of two linguistic devices: proper names or definite descriptions. Because Kuhn only deals with astronomical names it is unclear how his position applies with respect to descriptions. One might expect it to entail reference variance for descriptive singular reference. For descriptions contain general terms; and, according to Kuhn, general terms are subject to change of reference. Hence the reference of descriptions would vary with change in the extension of their contained predicates. Since names are uncommon in science, this problem may appear to outweigh Kuhn's concession of the stability of names.

However, this only appears to be the case if it is assumed that reference is determined by description. It will now be argued that, without the assumption of the description theory of reference, considerable scope exists for continuity and commonality of reference on the part of descriptions.

Because names are unusual in science descriptions require consideration as well. There is, however, an important class of descriptions which refer much as names do. In the case of descriptions which occur in referential use there may be continuity of reference to objects as there is with names.[18]

But let us first consider descriptions which occur in attributive use. A description may be used in such a way that its referent is what satisfies the contained predicate. The reference of descriptions in such attributive use depends on the description's being true of the object described. As a rule, a description is put to attributive use in a situation in which the causal and spatio-temporal relations operative in direct naming do not obtain. Consequently, their use in science is restricted to somewhat special contexts.

Attributive descriptions are directly affected by reference variance. Change in the extension of a predicate alters the reference of an attributive description containing the predicate. The earth, for example, is the referent of the contemporary description 'third planet from the sun'. But, on the geocentric use of 'planet' according to which the earth is not a planet, such a description would have referred to another body (perhaps Saturn).

But, given that reference change of general terms is limited on Kuhn's view, the scope for reference variance of attributives is also limited. Many general terms are stable in theory change, so the reference of attributive descriptions which contain them remains stable. Moreover, in those cases in which the extension of a predicate changes between theories, the same objects may be referred to by a predicate in both theories (as the moon is referred to first as a 'planet', later as a 'satellite'). Consider the case of the transfer of a member of a predicate's extension where the transferred member is the object referred to by the description. In such a case the old description no longer refers to the object, but another description containing the predicate to which the member has been transferred does refer to it.

Aside from these two options, the only other possibility of variance is a change where the purported reference of an older description is lost. That happens when a formerly posited entity is no longer thought to exist: e.g. when a hypothetical planet is found not to exist. This differs from the preceding case in which the reference of a predicate changes but the object to which it referred is referred to by another term. This third possibility appears to be restricted to ontological speculation in which empirical contact with the domain of investigation has yet to be fully developed.

I will now discuss descriptions which occur in referential use. Such use of descriptions is akin to direct ostensive reference. In order for a description to function referentially, the context of use as well as the intentions with which it is used determine reference. In the ordinary case, in which the description is true of the object to which the agent intends to refer, the reference of the description is multiply determined. But descriptions whose contained predicates are not satisfied by the intended referent can be employed referentially.

Thus, a geocentric thinker who uses the phrase "the body at the center of the cosmos" may count as saying something about the earth, even though nothing satisfies the description. To interpret the use of such an empty description as referential is not to take the description to be true of its referent. Rather, it credits the use of the description with being an act of reference to a particular object, in spite of its being misdescribed.

The description "the body at the center of the cosmos" has no contemporary use. At issue is whether in the past the description was used to make an assertion about what we refer to as the 'earth'. If the description occurred in referential use, the description may be replaced by an act of reconstructive semantics, putting the term 'earth' in its place. On such a reinterpretation of the phrase, the geocentrist may be understood to have mouthed a different sentence from the one he in fact did. The referring expression in his sentence is taken to be 'the earth' in recognition of the fact that the earth was the actual referent. There is no implication that the original description was true.

In such referential construals of descriptions containing empty predicates, there is discontinuity of reference at a purely linguistic level. For, considered in abstraction from the use to which it is put, the original description is not satisfied by any object. Since it is replaced by a putatively referring expression, there is a transition from an expression which fails to refer to one which purports to refer. However, given the referential use of the replaced description, there is continuity at the level of the act of reference. The discontinuity at the strictly linguistic level may be overlooked in favour of a sustained practice of referring to a common referent.

Descriptive singular reference constitutes a complexity with which Kuhn does not deal. Kuhn's stated views concern names and predicates rather than descriptions, so the preceding discussion is an extrapolation. Singular descriptive reference is a complication which yields little further ground for referential discontinuity. Admittedly, there is some scope for discontinuity on the part of attributives, but they are unusual. The only way to hold the general thesis that descriptive singular reference is discontinuous between theories would be to reject the referential use of descriptions and to take all descriptions as attributive. But that would conflict with Kuhn's partial endorsement of the causal theory.

In summary, the foregoing analysis reveals Kuhn's thesis to be a restricted doctrine of the referential variance of some terms. Revolutionary theory change alters the reference of a selected group of general terms, but leaves names and a large class of predicates referentially invariant. Since reference variance is restricted to selected kind terms, an overall stability of reference is maintained in the transition between theories.

Considered in broad terms, Kuhn's thesis that the reference of some theoretical terms is altered by change of theory is compatible with a modified causal theory of reference of the sort discussed in the final two sections of Chapter Two. In the first place, the causal theory accommodates reference change by taking into account the reference of term-tokens employed in particular acts of reference. In the second place, the potential for reference variance of theoretical terms is recognized by the causal-descriptive account of reference-fixing. Given that the causal role description necessary to fix theoretical term reference depends on theory, causal-descriptivism allows various possibilities for referential discontinuity. For example, it may turn out later that the reference of an earlier theoretical term was determined by a description which failed to secure a reference for it.[19] Or if the reference-fixing description associated with a theoretical term is changed, its extension may be partially altered in the manner of the higher-level categorial variance described by Kuhn. So, at a general level, the causal theory is consistent with Kuhn's view that the reference of some theoretical terms may change because of a change in the theory about the objects to which reference is made.

Kuhn's discussion of the problem of reference-fixing for theoretical terms is not specific enough to permit extensive criticism. However, there are a number of surface details on which comment should be made.

It was remarked earlier that Kuhn seems not to allow the possibility that the alloys were mistakenly thought to belong to the extension of 'compound' before Dalton. In fact, when he discusses change of reference Kuhn tends to employ a "success" vocabulary. For he says: "the alloys were compounds before Dalton, mixtures after"; and "the sun and moon were planets, the earth was not". The use of this linguistic device suggests that because the alloys were classified as compounds and the sun and moon as planets, the alloys were in fact compound and the sun and moon in fact planets. This amounts to taking classification behaviour as the factor which determines reference.[20]

To criticize Kuhn's use of a "success" vocabulary and his assumption that classification determines reference, the following brief remark will suffice. While it is true that reclassification may lead to extensional change, to construe all classificatory change as entailing extensional change overlooks a further possibility. Mistakes may be made in identifying the membership of a natural category. Thus while it might be decided by convention to no longer consider whales fish, equally it might be discovered that they were wrongly grouped with fish. In construing classificatory change on the model of the former, Kuhn runs the risk of losing sight of the latter.

As a related point, while Kuhn grants that categorial terms may be stable in theory-change, he fails to acknowledge the extent to which the reference of a theoretical kind term may be independent of any particular theory. While it is true that the reference of theoretical terms may be affected by theory-change, the fact that such a term's reference is constant need not depend upon the retention of the theory in which it is introduced. For example, the term's reference may be retained if the description originally fixing its referent is considered to have secured a referent.

Kuhn does not specify precisely how he thinks the reference of theoretical terms is fixed. However, his thesis that reference depends on classification seems to suggest that reference depends upon a specification of the kind to which an entity belongs. Vague as this is, it may still be objected that specification of the kind to which an entity belongs is not necessary for fixing reference.[21] While in some contexts it may be sufficient to specify kind, a specification of causal role may suffice as well. For example, one might succeed in referring to an unobservable entity by means of its causal role without correctly specifying the kind to which it belongs.[22] As against Kuhn, it should be noted that reference might survive change in classification provided that reference is originally fixed by specification of causal role.

It is of interest to consider the extent of referential variance which obtains between theories according to Kuhn's account. Kuhn, not surprisingly, does not specify the precise extent of change. The exact number of terms affected and the extent of the referential divide presumably depend on the particular theories in question. However, Kuhn does consistently express the thesis as a restricted thesis applying to a limited group of terms.[23] He stresses too that the terms involved in the change belong to a cluster of interrelated terms (see his 1983) and that the changes of reference are holistic in nature.[24] The picture that emerges is of a localized group of interrelated terms whose extensions are re-organized in the transition between theories.

To see what else characterizes the variant terms let us consider some of the examples Kuhn discusses. In the astronomical example the extensions of the terms 'planet' and 'star' change and the term 'satellite' is introduced. In the Dalton example, the invariant sets of salts, alloys, and metals are redistributed among the variant sets of compounds, mixtures, and elements. In the Newton-Einstein example, Kuhn talks of a change of the concepts of mass, force, space, and time.

There are two apparent features common to all the variant categorial terms: theoretical centrality and generality. They are terms with a central theoretical role in that they enter into the formulation of basic theoretical laws. Since basic theoretical laws are revised in fundamental change of theory, terms with such a central theoretical role are directly involved. Apart from centrality, such terms have a greater level of generality than the invariant terms. Terms such as 'compound', 'mass' and 'planet' represent more general categories than do, for example, 'alloy', 'physical body' and 'earth'.

So Kuhn's thesis is that reference change is confined to a central complex of higher-order categorial terms. It will now be argued, however, that the possibility of reference variance is not exclusively restricted to terms having these properties. An attempt will then be made to explain Kuhn's concern with such terms.

Having a high degree of generality seems not to be a necessary condition for a term to be subject to possible change of extension. For the extension of low-level kind terms such as 'alloy' could be altered in some way. Nor is generality itself necessary. A singular term introduced by an earlier theory may later turn out not to refer. Alternatively, singular reference may change by varying the object which the sigular term names.

More importantly, the metaphor of centrality breaks down under analysis. For, in any given domain, a number of different theories will ordinarily be applied as part of a single science. Thus there may be numerous classificatory concepts deriving from, and central to, a variety of different theories applied conjointly in the same domain.

In principle, there is no reason for terms employed in high-level theoretical laws not to remain unaltered while low-level empirical categories are altered. It would be possible to rearrange the members of the class made up of the metals and metal alloys in such a way that the extensions of the categories of compounds, mixtures and elements are all unaffected. The sort of categorial shift relevant to the reference change thesis can occur at any level within the network of theories applied in a domain.

This raises the question of why, given that reference change may occur among terms at any level of theory, there should be any special concern with terms in central theoretical laws. One possible defence of this concern which might be suggested is as follows. It might be said that in any scientific discipline there are a number of theories which apply at different levels and that each theory has its own central terms. For instance, there is a low-level theory about alloys and metals. It is independent of the theory about compounds, elements and mixtures. Change in the theory about alloys may affect the extension of its basic terms. But this defence effectively undermines the notion of centrality. For, depending on the context, any term would constitute a central term. On such an approach, centrality could not be the criterion for demarcating the terms of concern to Kuhn from those of none.

If this argument is right, then being a term centrally employed in the fundamental laws of theory does not distinguish terms susceptible to variation from those which are insusceptible. There would then be no point in insisting that categorial change is confined to a central theoretical level. For such change can occur across the board.

However, terms with a central theoretical role may be specified without requiring that they be intrinsically different from others. Kuhn's concern with central terms is due to his being mainly concerned with the deep structural changes of theory which are constitutive of scientific revolutions. Since theories affected by such revolutions are fundamental theories in their domain, the terms affected directly by the revolution are central theoretical terms in the basic laws. It is diagnostic of the terms to which Kuhn's reference change thesis applies that they are centrally implicated in a theoretical revolution; as such, they need have in their own right no unique propensity to change.

Kuhn's position may be further developed by placing more emphasis on the causal theory of reference and drawing a connection between centrality and Quinean unrevisability. Because central categorial terms figure importantly in fundamental theoretical laws they are comparatively unrevisable. For such laws are themselves protected from the impact of empirical falsifications upon outlying beliefs. Basic laws are revised only as part of change of the whole theoretical system.

In general, terms associated with the revisable theoretical periphery are those whose causal relations to their referents are the most direct. Terms for low-level kinds such as 'alloy', 'gold' and 'water' refer via direct causal relations. At this peripheral level there tends to be revision of belief rather than reference, though reference may change in local fashion without impact upon terms at a deeper theoretical level. Because of the immediacy of their referential links, moreover, profound conceptual change need not affect the reference of low level terms. This is consistent with Kuhn's view that terms for metals and alloys remain stable while higher-level terms (e.g. 'compound') undergo variance of extension. This application of the causal theory of reference provides an explanation, not otherwise to be found in Kuhn's account, of how there could be radical change of central concepts which leaves peripheral terms unchanged.

By contrast, terms in the central theoretical principles are least revisable and the links with their referents are more tenuous. Empirical problems impinging on the periphery can be deflected onto subordinate parts of theory without revision of basic laws. Revolutionary change of theory alters or replaces basic laws directly affecting the terms they contain. It is because of this that radical change of theory affects these terms in particular. Furthermore, to the extent that they are disconnected from their referents, such terms are more prone to extensional variance. For the reference of the more central terms may depend entirely on theoretical description.

Notes

1. It might be objected that pragmatic conditions may be invariant even if objects were theory-relative and not shared by theories. Yet if the pragmatic conditions really are invariant, that is enough to ensure the objects are invariant. The theory-relativity of objects will be criticized in section 7.5.
2. Cf. "the meaning of every term we use depends upon the theoretical context in which it occurs. Words do not "mean" something in isolation, they obtain their meanings by being part of a theoretical system" (Feyerabend 1965, p. 180).
3. Feyerabend argues that the meaning of colour vocabulary varies with theory in his (1981c, p. 29).
4. This pattern of inference is Feyerabend's standard procedure. For example, he also applies it to Einsteinian and Newtonian concepts of mass (1965, pp. 168-9). The form of argument derives explicitly from Nagel's conception of reduction, of which Feyerabend's argument is a criticism.
5. Nagel (1960, p. 302), quoted in Feyerabend (1981d, p. 67).
6. This analysis gains support from the following explication of co-extensiveness which Feyerabend gives in his (1965, p. 184): ""X" and "Y" must possess the same extension, which means the intension of "X" must not contain components denied in the intension of "Y" or vice versa."

7. It will be argued in 6.3 that Kuhn's talk of "world-change" should not be taken literally.
8. Alternatively, it is compatible with the existence of two physical quantities, relativistic as opposed to proper mass, between which the reference of Newton's tokens of 'mass' was indeterminate. See Field (1973, pp. 465-70). However, as Field shows, "Newtonian mass" cannot be either because it would have properties in common with each as well as some which neither has.
9. E.g. Putnam interprets Kuhn in this manner when he says: "Kuhn talks as if each theory does refer — namely, to its own 'world' of entities" (Putnam, 1978, p. 23).
10. Non-realist interpretations of Kuhn are dealt with in Chapters 6 and 7.
11. There is an apparent slip of the pen: where he says "universe of variables" he presumably means "universe of objects over which the variables of quantification range".
12. For such a realist treatment of the case of 'mass', see Musgrave (1979, p. 344), Nola (1980a, p. 334) and Smith (1981, pp. 119-24). Nola (1980a, pp. 342-6) uses the causal theory of reference to argue for the stability of reference of 'mass'.
13. See Quine (1960, pp. 29ff).
14. I will set aside the historical question of whether in fact the reference of 'compound' did change. A number of writers have protested against Kuhn's interpretation of the case. Thus Nola says that "it does seem clear that the reference of 'compound' is not what changes; rather Dalton got us to change our beliefs about what we regarded as falling within the extension of 'compound'" (1980b, p. 510). Cf. Fine (1975, p. 21).
15. The point is reminiscent of the "qua problem" discussed in 2.6: viz. that the strictly causal or physical relationship involved in ostension does not determine which kind is the intended referent.
16. Cf. Fine (1975, p. 21) for this point.
17. The idea that Kuhn's thesis of reference change is restricted to high-level kinds is found in Hacking (1983, p. 110). Hacking's interpretation of Kuhn as a nominalist will be discussed in 7.2.
18. For Donnellan's distinction between referential and attributive uses of definite descriptions, see section 2.4.

19. Of course, this is not strictly a change of reference, but a discovery of reference failure.
20. As Fine notes: "The classification behaviour, then, becomes the living embodiment of the concept. As it classifies, so does the concept pick out the reference" (1975, p. 21).
21. For analysis of the issue see the discussion of Enç and Nola in section 2.6.
22. E.g. Nola points out that reference was made to oxygen before it was known to be an element (1980b, p. 525).
23. E.g. "what characterizes revolutions is ... change in several of the taxonomic categories" (1981, p. 25), "successive theories are incommensurable ... in the sense that the referents of some of the terms which occur in both are a function of the theory within which those terms appear" (1979, p. 416).
24. E.g. "redistribution [of objects among taxonomic categories] always involves more than one category and since those categories are interdefined, this sort of alteration is necessarily holistic" (1981, p. 25).

6 Against the idealist interpretation

6.1 Introduction

The next two chapters address broadly metaphysical issues associated with the radical reference change thesis. The problem of reference change raises questions about the relation between theory and reality. At the hands of Kuhn and Feyerabend talk of reference change often suggests that in some sense reality depends on theory.

On occasion, the way they express their views suggests that the incommensurability thesis is an idealist rejection of a world independent of theory.[1] For, if taken literally, the way they sometimes treat reference and the suggestion that theories are about their own worlds credits theory with control over existence.

That the impression of idealism is widespread is attested by the number of writers who take incommensurability as an issue between realism and some form of idealism. Putnam argues that Feyerabend's view about reference rests on an "idealist or idealist-tending world view" (1975a, pp. 196-9, 207). Boyd (1984) defends realism against "constructivist antirealism" which he associates with Kuhn and incommensurability. Devitt takes Kuhn and Feyerabend to espouse "relativistic weak [or "fig leaf"] realism", according to which "scientific theories are imposed on things-in-themselves to yield an ontology-relative-to-that-imposition" (1984, p. 139). For Scheffler, Kuhn's view of science presents a "bleak picture, representing an extravagant idealism", which is a "reductio ad absurdum of the reasoning from which it flows" (1967, p. 19).[2]

It will be argued in this chapter that the extreme interpretation of Kuhn and Feyerabend as idealists who deny the existence of an independent reality is incorrect. Given that the incommensurability thesis is consistent with there being a world independent of theory, it is unnecessary to argue against such extreme idealism here.

There would, in any case, be little to add to Scheffler's point that "extravagant idealism" is absurd. That the world itself should change with shift of theory may be dismissed as contrary to experience. Besides, if Kuhn and Feyerabend were extreme idealists their philosophy of science would risk incoherence. For on both of their views of science a role is played by empirical problems which derive from experience and observation. Such problems imply the possibility of counter-evidence and failed prediction. But if the world itself is determined by the theory we accept, no sense can be given to the idea of a recalcitrant experience. For if experience can fail to accord with a theory, then at least something in the world is not produced by the theory.

To grant that reality is independent of theory is, however, consistent with "constructivist" forms of idealism which accept a theory-independent reality. According to constructivism, both language and theory refer to a constructed world which is in part produced by human cognition. The issues raised by constructivism will be considered in the following chapter.

The rest of this chapter is organized into three parts. In 6.2 the idealist interpretation is introduced and the requisite notion of idealism is discussed. It is argued in 6.3 that Kuhn and Feyerabend are committed to a unique reality which is invariant between theories; hence the "world-change" image is to be taken figuratively. Section 6.4 discusses the extent to which their treatment of reference is idealist.

6.2 The idealist interpretation

Kuhn and Feyerabend do not explicitly argue for idealism and tend to make realist claims.[3] Thus the charge of idealism is based on an interpretation of idealistic aspects of their position and its formulation.

The main basis for the charge of idealism is Kuhn's use of the "world-change" image. His (1970a) contains numerous claims that the world changes with change of paradigm. Such claims are accompanied by images of entities coming into existence and observers of the same domain seeing different entities. For example, "pendulums were brought into existence by something very like a paradigm-induced gestalt switch" (1970a, p. 120), "Lavoisier... saw oxygen where Priestley had seen dephlogisticated air" (p. 118), "we have had to alter the fundamental structural elements of which the universe to which [Newton's laws] apply is composed" (p. 102).

On occasion Feyerabend employs a similar idiom. The following represents his most extreme formulation of the world-change idea:

> we certainly cannot assume that two incommensurable theories deal with one and the same objective state of affairs (to make the assumption we would have to assume that both at least refer to the same objective situation. But how can we assert that 'they both' refer to the same situation when 'they both' never make sense together? ...) Hence, unless we want to assume that they deal with nothing at all we must admit that they deal with different worlds and that the change (from one world to another) has been brought about by a switch from one theory to another... we no longer assume an objective world that remains unaffected by our epistemic activities, except when moving within the confines of a particular point of view. We concede that our epistemic activities may have a decisive influence even upon the most solid piece of cosmological furniture — they may make gods disappear and replace them by heaps of atoms in empty space. (1978, p. 70)

This passage is uncharacteristic, since Feyerabend's talk of "worlds" is usually less amenable to an idealist reading.[4]

In addition to the world-change image, both writers occasionally discuss reference change in a way which suggests idealism. They describe reference change as if actually existing things in the world themselves change along with theory. Thus Kuhn, for example, writes that "alloys were compounds before Dalton, mixtures after" (1970b, p. 269). Elsewhere he says:

> Before [the transition from Ptolemaic to Copernican astronomy] occurred, the sun and moon were planets, the

earth was not. After it, the earth was a planet, like Mars and Jupiter; the sun was a star; and the moon was a new sort of body, a satellite. (1981, p. 2)

On a literal reading of either statement something in the world has changed its nature. If alloys were compound then became mixtures, at the very least that means that the alloys themselves underwent a change.

Sometimes Kuhn and Feyerabend describe conceptual difference as difference of entity. It is as if defining a concept within a theory brings corresponding entities into existence, and different theories bring into being different entities. Both speak of Newtonian or classical as opposed to Einsteinian or relativistic mass.[5] To assume that there is either classical or relativistic mass means that there is something of which Newton's or Einstein's physics is true. But since the two theories contradict each other about the nature of mass, the only way for both to be true is if mass itself varies with theory, which implies that reality itself is transformed by the change of theory.

Such intimations of idealism can appear as bold idealistic pronouncements when placed in the context of their overall philosophy of science. Their arguments for the theory-ladenness of observation and the resistance of theory to empirical refutation serve to minimize the impact that the world can have on theory. They both argue against a sharp distinction between fact and theory.[6] Taken to the extreme, rejection of the fact-theory dichotomy suggests that facts depend on theories and that there is no independent world of objective facts. Both writers are dismissive about truth. Combining their views about the determination of meaning by theory, theory-ladenness of observation, radical ontological changes, and antagonism to truth and facts, there seems to be no place left for the real world. On the overall picture which emerges from their philosophy of science it is unclear whether science has anything to do with a theory-independent reality at all.

There is a difference, of course, between conceiving of the world as a product of theory and showing that theories have a certain immunity to refutation by experience. The world-change image and theory-relative reference suggest that reality depends on theory and fluctuates relative to it. But an approach to science on which theories are not rigorously controlled by an independent realm of fact need not be one on which the objective world does not exist. Such an approach is consistent with the existence of an independent reality. One on which the world varies with theory is not.

In the sense in which Kuhn and Feyerabend seem idealist, reality varies relative to theory because what is real depends on theory. On this conception, incommensurable theories bring completely distinct worlds into being. Since there is no unique world and incommensurable theories create their own world, the transition between such theories is a transfer from one world to another. There can be no question, then, of comparing such theories as rival views of a common world, for the worlds that they are theories of are not the same. To choose a theory over one with which it is incommensurable is not a choice of which theory best deals with this world, but rather a choice of reality.

To show that Kuhn and Feyerabend are not idealists in this sense it suffices to show that they are committed to an independent world which is invariant between theories. The argument of the next two sections attempts to establish such a commitment.

One qualification about this approach should be stated, however. There is a form of idealism according to which reality is independent of theory. This is the view that reality is itself mental or is produced by the mind, but is not affected by variation of belief or theory. Commitment to a theory-independent reality does not disprove the charge of idealism if idealism is taken in this sense. For it can always be maintained that the reality which is invariant relative to particular theories, or even theories in general, is still somehow manufactured by the mind. Yet without an explicit endorsement of such an extreme mentalism, it cannot be justifiably attributed. And even if Kuhn and Feyerabend were idealists in this sense, it would still not affect the issue of incommensurability. For if the world is invariant relative to theory, theories may be brought to bear on and compared with respect to a world independent of theory.

Of greater relevance are the constructivist views dealt with in the next chapter. Constructivism makes the minimal realist concession that there is a reality whose existence and properties are independent of theory. According to constructivism, however, such a reality is inaccessible; the reality which we experience is a constructed one. The only reality relevant to cognition is a world produced by cognition itself. The existence and properties of that reality depend upon our theories and concepts and may vary as they change. Because of the minimal realist concession of a reality not dependent upon theory, constructivist readings of

Kuhn and Feyerabend cannot be dismissed by the arguments of the next two sections.

6.3 One world is enough

Kuhn and Feyerabend both employ the analogy of gestalt figures to illuminate the relationship between incommensurable theories. In viewing a gestalt diagram there are normally two visual images to be seen, the perception of both of which cannot be concurrent. Similarly, incommensurable theories present points of view which cannot simultaneously be held in mind. Thus the switch between the "worlds" of incommensurable theories is mediated at the psychological level by a shift of gestalt.

Dilworth (1981, p. 91) points out that to apply the analogy to the case of alternative theoretical points of view requires that incommensurable theories be alternative perspectives on a common point of focus. If not, the analogy breaks down. If it is meant to be a strict analogy, incommensurable theories must be alternative theories of the same world, which is independent of the theories in question.

At a number of places Feyerabend has sought to clarify his view about the relationship between theory and reality. It is a minimal realist conception. At one point he comments as follows:

> When we go from classical physics to relativity, what remains the same are the objects. The objects are what they are, only we think different things about them.[7]

This straightforwardly accepts that alternative theories may refer to an objective reality not dependent on either theory. In fact, to say "the objects are what they are" suggests something stronger: viz. that the objects exist in their own right independent of any theory. Elsewhere he further stresses his commitment to that view in response to an objection:

> nor do I ever say that what is is the same as what is thought to be ('conflation' of theoretical object and real object). 'Realism'... does not mean that the world is identified with the theoretical object, it means that one tries to understand the real in theoretical terms rather than regarding it as 'given'. (1978, p. 171)

Together with the gestalt analogy, these quotations are indicative of a realist attitude to the relation between theory and reality.

In brief, theoretical changes are changes which occur at the level of theory, not alterations of the constituents of reality.

More decisive are considerations about the language Feyerabend employs in the course of a discussion of an incommensurable theory pair. Examination of the discussion reveals that reference is made to theory-neutral objects. This point is cogently argued by Kordig, who shows that Feyerabend employs a description which can be used as a neutral metalinguistic specification of a common area of reference for a pair of theories. From this Kordig concludes that there is "(extensional) meaning invariance" between the theories (1971, p. 95).

Kordig draws attention to Feyerabend's description of the subject matter of Galilean physics. Feyerabend says that Galilean physics is about:

> the motion of material objects (falling stones, penduli, balls on an inclined plane) near the surface of the earth. (Feyerabend, 1981d, p. 57)

Kordig then notes that this description also applies to the subject matter of Newton's terrestrial physics. The description may therefore be used as a metalinguistic specification of the joint domain of reference for the two theories. Kordig formulates such a specification as follows:

> Using Feyerabend's own words, we can describe the neutral observational objects in terms which are neutral to T_1 [Galilean physics] and T_2 [Newtonian physics]: Both T_1 and T_2 refer to material objects such as falling stones, penduli, balls on inclined planes, etc., each of which are near the surface of the earth... The terms in [this] description occur in our (and Feyerabend's) meta-language in which we talk about T_1 and T_2. (1971, pp. 95-6)

Feyerabend's use of a neutral description which refers to the objects investigated by the two theories implies that, whatever the relation between their conceptual apparatus, they are about a fixed and independent reality.

We can also apply Kordig's point to Feyerabend's discussion of impetus. Feyerabend contrasts the account given of inertial motion in the impetus theory with that given by Newton. In the following quote from Feyerabend the phrase 'a moving object in empty space' functions in a theory-neutral manner to pick out a class of motions which both theories purport to describe:

> A moving object which is situated in empty space and which is influenced neither by gravity nor by friction is not outside the reach of any force. It is pushed along by the impetus, which may be pictured as a kind of inner principle of motion ... We now turn to Newton's celestial mechanics and the description in terms of this theory, of the movement of an object in empty space ... Quantitatively, the same movement results. (1981d, p. 65)

The contrast would be pointless if the two theories were not supposed to be about the same thing: what is contrasted is alternative views of inertial motion. The phrases "a moving object" and "the movement of an object" describe the theory-independent entities referred to by the two theories. This point is reinforced by the claim in the last sentence that the theories agree in predicted quantitative value, which can only be asserted on the assumption that an identical motion is measured. Thus these two phrases could also be used as neutral metalinguistic specifications of reference for the terminologies of the two theories.

A similar neutral description is found in the same discussion. Feyerabend says that "in the Aristotelian theory, the natural state in which an object remains without the assistance of any causes is the state of rest" (1981d, p. 65). This differs from Newtonian physics, in which "it is the state of being at rest or in uniform motion which is regarded as the natural state" (p. 65). This is a comparison of two rival conceptions of the natural state of bodies, which directly implies that the theories in question are about the same objects.

Feyerabend concedes that common referents of incommensurable theories can be described in a joint metalanguage. This concession occurs in the context of a discussion of the possibility of crucial experiments between incommensurable theories. Feyerabend mentions experiments which disconfirm classical mechanics while confirming relativity theory. The puzzle is how such experiments could be relevant to both theories if, qua incommensurable theories, they do not possess common meaning. He grants that an experiment can be so described as to make it clear that the very same experiment is relevant to both theories. Then he claims:

> the identification [of the crucial test] is ... not contrary to my thesis, for we are not using the terms of either relativity or of classical physics, as is done in a test, but are referring to

> them and their relation to the physical world. The language in which this discourse is carried out can be classical, or relativistic, or Voodoo. (1975, pp. 282-3)

To refer to terms and their relation to the world is to speak about such terms in a metalanguage. The point is that the identity of a crucial experiment can be asserted in a metalanguage even if the theories are incommensurable. But the existence of shared experiments implies the existence of objects independent of either theory and that such objects may constitute common referents.

There is a separate element of Feyerabend's philosophical position which commits him to an independent world. Feyerabend's pragmatic theory of observation (see 5.2) implies continuity of relation to a stable extralinguistic reality between incommensurable theories. For according to the pragmatic theory, the use of an observation sentence is acquired by a conditioning process in the context of the relevant physical surroundings. Though the meaning of an observation sentence may vary with theoretical context, the pragmatic conditions of its use may remain constant. But since the conditions of such use constitute the physical surroundings in which it is appropriately employed, and since they may remain stable, it follows that the world itself is not affected by change of meaning induced by change of theory.

Kuhn presents a slightly different case. With Feyerabend it is a matter of finding an assurance that incommensurable theories are about a shared independent reality. With Kuhn it is a matter of assuring ourselves that his talk of "world-changes" is not meant literally.

We may apply Kordig's point about use of a metalanguage to Kuhn as well. To take just one example, in the following quote joint reference is singled out for specific terms of two different theories using a neutral term:

> I am, for example, acutely aware of the difficulties created by saying that when Aristotle and Galileo looked at swinging stones, the first saw constrained fall, the second a pendulum. (1970a, p. 121)

If we assume that 'see as' implies 'describe as', this passage entails the following:[8]

> Aristotle and Galileo observed swinging stones.
> Aristotle described swinging stones as constrained fall.
> Galileo described swinging stones as penduli.

Analyzed in this way, Kuhn's remark is couched in a meta-language which contains non-synonymous object-linguistic expressions and a neutral term which describes their common reference. Though the remark itself is not explicitly formulated in such a way that both terms and referents are referred to, the term 'swinging stone' neutrally designates a phenomenon which Galileo and Aristotle understand quite differently, and for which they employ distinct vocabulary. So the expression 'swinging stone' may be used to metalinguistically specify the reference of Aristotle's and Galileo's expressions.

There are compelling reasons for taking the "world-change" idiom as a metaphor. First, Kuhn's talk of world-change tends to be heavily qualified. For instance, in Section 5.3 we took note of Kuhn's remark that in paradigm change it is "rather as if the professional community had been suddenly transported to another planet" (1970a, p. 111). The "as if" clause has the effect of removing the phrase's literal assertoric force. Moreover, the very next sentence states a proviso which deprives it of its literal meaning: "Of course, nothing of quite that sort does occur".[9] And in the next one he specifies the constrained sense in which the metaphor is meant: it is the way "scientists ... see the world of their research-engagement" that changes.

It is not merely that Kuhn hedges his use of the image. He issues warnings and disclaimers as well. In his (1970a) he cautions against taking it literally. He was having difficulty expressing a point:

> In a sense that I am unable to explicate further, the proponents of competing paradigms practice their trades in different worlds. (1970a, p. 150)

Subsequently, he completely repudiated the literal "world-change" position. At the time of writing the 'Postscript' to his (1970a) he began to talk of "stimuli", which were conceived as objective when compared with the subjective "sensations" which form the content of our sensory experience. He said "We posit the existence of stimuli to explain our perception of the world, and we posit their immutability to avoid both individual and social solipsism" (1970a, p. 193). A more forthright disavowal occurs elsewhere:

> In *The Structure Of Scientific Revolutions* ... I repeatedly insist that members of different scientific communities live in different worlds and that scientific revolutions change the

world in which a scientist works. I would now want to say that members of different communities are presented with different data by the same stimuli. Notice, however, that that change does not make phrases like 'a different world' inappropriate. The given world, whether everyday or scientific, is not a world of stimuli. (1977, p. 309, fn. 18)

Kuhn's disclaimers and reservations show that he does not endorse a radical "world-change" thesis, whatever the image may suggest. Consideration of the context in which Kuhn employs the image reveals, furthermore, that the extreme reading of the "world-change" image is not supported by the text.

A number of writers have noted that Kuhn assumes an independent reality throughout his work, even in the passages which seem to deny it.[10] Where Kuhn speaks of the world as something possessed by a particular theory, he often uses another expression, such as 'nature' or 'the environment', to refer to a fixed and independent reality.

Brown, for example, draws attention to the following passage in Kuhn (1970a):

> The subject of a gestalt demonstration knows that his perception has shifted because he can make it shift back and forth repeatedly while he holds the same book or piece of paper in his hands. Aware that nothing in his environment has changed, he directs his attention increasingly not to the figure (duck or rabbit) but to the lines on the paper he is looking at. Ultimately he may even learn to see those lines without seeing either of the figures... (Kuhn, 1970a, p. 114)

Brown then points out that:

> Kuhn's talk of 'the world changing' as a result of the shift between incommensurable frameworks must not be read as asserting that nothing remains constant. There has been much unnecessary misunderstanding generated by Kuhn's use of the term 'world' in this context. It was not a felicitous choice of terminology, but a careful reading of Kuhn's remarks, including the passage quoted above, makes it clear that Kuhn is working with a distinction between 'the world' and 'the environment' throughout this discussion, with the latter term serving to designate entities which exist quite apart from our theories, frameworks, or paradigms, and which scientific research is always ultimately concerned with. (Brown, 1983a, pp. 19-20)

In accordance with this, we find Kuhn saying that "the world that the student enters is not ... fixed ... by the nature of the environment ... and of science" (1970a, pp. 111-2), which distinguishes between the changeable world experienced by the scientist and environing reality.[11] At another place, Kuhn speaks of "the world determined jointly by nature and by the paradigm upon which Galileo ... [was] raised" (p. 125). It would be perverse to think that this implies a strange determination of reality by paradigm, for the term 'nature' has the sense of a fixed reality and 'world' the sense of a constructed and experienced one.

Finally, Kuhn's notion of an anomaly depends on the existence of a world which is independent of theory. In Kuhn's sense an anomaly is a problematic empirical phenomenon not satisfactorily accommodated within the theoretical framework of a paradigm. One example Kuhn gives of an anomaly is Roentgen's discovery of x-rays. Kuhn tells how, in the course of work on cathode rays, Roentgen noticed a peculiar glow on a screen situated at some distance from his shielded apparatus. He describes what ensued as a "perception of anomaly — of a phenomenon, that is, for which his paradigm had not readied the investigator" (1970a, p. 57).

But phenomena which are encountered by accident or which are unanticipated by a paradigm cannot be artifacts of the paradigm. Since such phenomena are not foreseen within the paradigm, it is rather a case of being confronted with unexpected occurrences by extra-theoretic reality. The only way to explain the possibility of empirical anomalies is on the assumption of a theory-independent reality whose finer points resist discovery.

Moreover, a paradigm which is accepted after a scientific revolution solves the anomalies which precipitated the revolution. Anomalies must, then, have a modicum of independence from paradigms if they are capable of relations to different paradigms. That would be impossible if paradigms did not provide competing explanations of the same world.

6.4 Idealism and reference change

On occasion, Kuhn and Feyerabend discuss changes of classificatory scheme in a manner which suggests a change in the nature of the items classified. Thus they say that alloys were compounds, then mixtures; that the sun was a planet, then a star; that Newton had a theory about classical mass, Einstein about relativistic mass.

This manner of speaking imputes change to what is classified rather than to the classificational system. What is referred to is said to have been one sort of thing and then to have become another sort of thing. This suggests that the objects and magnitudes are themselves transformed. Now, a class of objects cannot first be one kind of thing, then become a distinct kind, and continue throughout to be the same kind of thing. Alloys cannot be both compounds and mixtures, the Sun cannot be both a planet and a star, and mass cannot be invariant and relative. To say that the alloys were transformed from compounds into mixtures means that the alloys themselves underwent a change of structure. But if change of classification leads to a change in the very stuff of which alloys are made, then that is idealism.

What would follow about reference if there were such a transformation of the world? Given the purported change in the nature of the alloys, reference would change too. The extensions of 'compound' and 'mixture' change if the alloys move from the class of compounds to the class of mixtures. And if alloys actually change their physical structure, there is a sense in which 'alloy' must change its reference as well. For in the one case 'alloy' refers to a kind which is chemically compound and in the other to a mixture; so the natural kind to which 'alloy' refers changes.

If there were a comprehensive change of reference such that everything referred to changes its nature in the above fashion, then the world itself would change. This is the position that emerges when the above way of talking about reference change is combined with Kuhn's world-change metaphor.

Such change in the nature of objects would not, however, be less idealist if it failed to be fully general. If only the constitution of the alloys were to change while all else remained the same, the way the transmutation occurs as the result of a change of theory is still idealist.

Thus the mark of this type of idealism is not extensional disjointness due to reference to distinct worlds. What characterizes the idealism implied by such changes of reference is that reference succeeds before and after changes of theory. In the above cases, whether or not the reference change is comprehensive, the class terms are alleged to succeed in picking out different classes: the alloys *were* compounds and the Sun *was* a planet. That theories should design their own referents in this way constitutes the idealism inherent in this way of handling reference.

Feyerabend's view of radical change of reference (see 5.2) lacks this crucial idealist assumption of successful reference. On his view the languages of incommensurable theories are extensionally disjoint, but not because they actually succeed in referring to distinct worlds.

For Feyerabend, incommensurable theories associate mutually exclusive conceptual or descriptive content with their terms. In 5.2 it was shown that Feyerabend assumes that descriptive content determines reference in the manner of a description theory of reference. Given that assumption about reference determination, since incompatible descriptions are not jointly satisfiable, it follows that the languages of incommensurable theories are extensionally disjoint.

Feyerabend does not construe this total failure of co-reference as successful reference to distinct worlds; rather, at least one of the theories fails to refer. Instead of being idealist, the claim of failure rather than success of reference makes his position a fundamentally realist one.

In view of Feyerabend's commitment to a world invariant between theories (6.3), idealist-sounding remarks of his may be interpreted as a manner of speaking. At a number of places, for example, Feyerabend uses the locutions 'classical mass' and 'relativistic mass': e.g. "the values obtained on measurement of the classical mass and of the relativistic mass will agree in the domain" (1981d, p. 81, cf. 1965, p. 168). Such locutions suggest the existence of different sorts of mass: classical physics refers to classical masses, whereas relativistic physics is about something different, relativistic mass. But if we assume the world remains fixed, the passage can be read as elliptical for: "the values obtained on measurement of the mass understood in accordance with the classical concept thereof and with the relativistic concept will agree in the domain". Moreover, it is clear in the context that Feyerabend is using the locutions as shorthand; i.e. 'relativistic mass' means 'the relativistic concept of mass'.

To take Feyerabend in this way as a realist with a description theory of reference may seem irreconcilable with the idealistic passage quoted in 6.2. For there Feyerabend explicitly states that:

> unless we want to assume that [incommensurable theories] deal with nothing at all we must admit that they deal with

> different worlds and that the change (from one world to another) has been brought about by a switch from one theory to another... (1978, p. 70)

But since Feyerabend is committed to a world independent of theory (as argued in 6.3), the idealistic tone of this remark should be read as exaggeration.

Moreover, what leads him to the idealist manner of speaking is a false dichotomy between referring to nothing at all and dealing with different worlds. For immediately prior to the above remark he claims that:

> we certainly cannot assume that two incommensurable theories deal with one and the same objective state of affairs (to make the assumption we would have to assume that both at least refer to the same objective situation. But how can we assert that 'they both' refer to the same situation when 'they both' never make sense together? ...) (1978, p. 70)

Feyerabend is led to the idealist conclusion that incommensurable theories refer to different worlds because such theories "never make sense together". Now, to say that theories fail to "make sense together" is a shorthand way of describing the relation between incommensurable theories. For theories are incommensurable if "the conditions of concept formation in one theory forbid the formation of the basic concepts of the other" (1978, p. 68). But to conclude from the mutual exclusivity of their concepts that theories cannot co-refer is to assume that descriptive content determines reference. So Feyerabend is led into idealism by the assumption, licensed by the description theory of reference, that incommensurable theories can have no common reference. Given his commitment to an extra-theoretic world, such idealism seems an exaggeration to which he is needlessly drawn by a mistaken theory of reference.

Kuhn's early position (5.3) is a more suitable candidate for idealism. When he says "the physical referents of these Einsteinian concepts [e.g. mass] are by no means identical with those of the Newtonian concepts that bear the same name" (1970a, p. 102), he seems to be saying something idealist about reference.

The claim suggests that the Newtonian and Einsteinian concepts successfully refer to completely distinct sets of things. That is, there is no single set of material objects which serves as a shared domain for these concepts of mass. Rather, in moving from the one conceptual apparatus to the other a world with one type of mass is replaced by a world with the other type of mass. This suggestion provides genuine textual support for the charge of idealism.

There is no need to conclude from this that Kuhn is an idealist. At most, it must be conceded that there is a textual basis for the charge. For, in light of Kuhn's commitment to an independent world (see 6.3), the hint of idealism may be taken as misstatement. This interpretation is further supported by his subsequent clarifications and disavowals which show him not to be an idealist.[12] Though the above passage is amenable to an idealist reading, it is not indicative of a general commitment to idealism.

The significance of the textual basis for the charge is even less than this suggests. For it was shown in 5.3 that Kuhn's original treatment of reference in his (1970a) was triply ambiguous. Aside from the idealist reading, the above passage is also consistent with his later view that there is a limited classificational change (to be dealt with momentarily). More plausibly, it may be taken in the way we have interpreted Feyerabend. On such an interpretation, Kuhn's term 'referent' in the above passage is elliptical for 'putative referent'. What changes is not mass but the description of what is purported to constitute mass. Since the latter analysis does not imply that the world changes with change of concept, it is not idealist.

Kuhn's later position is that a restricted conceptual upheaval characterizes the transition between incommensurable theories (5.4). This view may encourage thoughts of a more limited form of idealism. It will be helpful to introduce a distinction between global and local idealism. Global idealism is the view that there is a massive transformation in reality because of a theory change. This global sense is suggested by the world-change metaphor. Local idealism is the view that the world as a whole is unaltered but that isolated changes occur here and there.

On Kuhn's later view the effect of conceptual change upon the language in which incommensurable theories are couched is limited to a few central terms. This leaves a common stock of semantically invariant terms jointly available to both theories. It follows from the existence of a shared semantically neutral language that no radical displacement of domain of reference takes place in the transition between incommensurable theories.

The question of idealism may be raised with respect to the limited area of language affected in theory change. When a theory replaces an incommensurable predecessor, it imposes a novel categorial structure on the world via the conceptually transformed portion of the language. It might be suggested that the world does not in its own right possess the relevant structure independently of cognition. On the local idealist construal of this suggestion, mental intervention brings about the requisite changes in the structure of the world. In contrast, the constructivist construal is either that there is no fact of the matter about such structure, or that there is but it is epistemically inaccessible; in either case structure is a matter of convention.

In 7.2 the relevant constructivist position, a form of nominalism, will be discussed. For now it suffices to show why Kuhn is not a local idealist. Consider what the local idealist must say about categorial change. On an idealist view, the existence of such categories must depend on a theory which brings them into existence. This is not merely to say that the classificatory system is brought into being. Rather, the actual divisions between kinds of things are brought into existence or transformed by the process of classification.

Now, in the case of the alloys this implies that the alloys changed their compositional structure by shifting from the category of the compounds to that of mixtures. Such a change does not represent an artifical classification or a fallible attempt to determine what categories exist. Rather, it implies that a change of classification directly affects the composition of material objects. But such an idealist implication of the power of mind over matter is inconsistent with Kuhn's acceptance of a reality whose existence and properties are independent of theory.

Notes

1. Hacking notes that "it is extremely easy for a reader of *Structure* to think that its author is, *au fond*, a raving idealist" (1979, p. 229), an interpretation he rejects. For further discussion of idealism in connection with incommensurability see Musgrave (1979, pp. 337-8), Nola (1980a) and Rorty (1980, pp. 273-7).
2. Cf. Suppe (1977, p. 151).
3. Typically difficult to interpret is Kuhn's claim to be an "unregenerate realist" (1979, p. 415), for he confesses to being "uneasy" about "one real world, still unknown but toward which science proceeds by successive approximation"

(p. 418), and suggests in constructivist vein that "what we refer to as "the world" [is] perhaps a product of mental accommodation between experience and language" (p. 418).

4. Cf. "... the modes of representation used during the early archaic period in Greece ... give a faithful account of what are felt, seen, thought to be fundamental features of the world of archaic man" (1975, p. 248); "why should the perceptual world of the ancient Greeks coincide with ours?" (1975, p. 249).
5. E.g. Kuhn (1970a, p. 102), Feyerabend (1981d, p. 81) and (1965, p. 169).
6. See Feyerabend (1965, pp. 174f), (1975, p. 38); Kuhn (1970a, pp. 52ff).
7. Comment in discussion in Hanson (1970, p. 247), cited in Dilworth (1981, p. 91).
8. We may disregard as irrelevant that neither Galileo nor Aristotle spoke English.
9. A glance at other passages (e.g., 1970a, pp. 117-8, 121, 150) reveals that such remarks are nearly always hedged and expressed in a tentative tone.
10. E.g. Brown (1983a, pp. 19-20) and (1983b, p. 97), Devitt (1984, p. 137), Mandelbaum (1982, pp. 50-2).
11. Cf. Brown (1983b, p. 97).
12. For the clarifications see the discussion in 5.4; for the disavowals see 6.3.

7 Against constructivism

7.1 Introduction

This chapter criticizes a number of related doctrines collected under the head of "constructivism". The basic aim of the chapter is to defend the view that incommensurability is a problem of language rather than metaphysics.

Chapter Six has shown that the incommensurability thesis is not an idealist thesis of change of world with change of theory. That leaves the possibility, however, that there is a weaker sense in which the "world" of a theory is not the real world.

According to constructivism, the objects, kinds and even the "world" to which a theory refers are not independent of theory. In some sense they are a product of theory. What unites the various constructivist doctrines under one heading are two characteristic theses. As opposed to idealism, constructivism grants that there is a reality whose existence and character are independent of mental activity. Yet constructivism denies that the world dealt with by a scientific theory is such a mind-independent reality itself.

If this were right, the problem of incommensurability would not be essentially a problem in the theory of reference. This would undermine the referential approach to theory comparison. For the approach to comparison by means of co-reference assumes that different conceptually variant theories may be applied to a common world.

If theories make their own entities, the problem of how theories may co-refer to theory-independent entities arises in an extreme form. For the sense in which theories which construct their own entities could be about the same world is quite unclear. So constructivism provides the basis for a thesis of the radical incomparability of theories.

The chapter divides into sections each dealing with a distinct constructivist position. Section 7.2 considers the idea that Kuhn is a nominalist about high-level natural kinds. In 7.3 an argument for conceptual relativism is discussed. Section 7.4 considers the thesis that theories deal with different worlds because of the theory-dependence of observation. In 7.5 the relativity of objects is considered. And 7.6 is about the rejection of reference.

7.2 Nominalism

According to Kuhn's mature position, a localized conceptual change occurs in the transition between incommensurable paradigms. The question arises of the status of the categories or kinds which systems of concepts are about. Do categories exist independently or do they depend on an act of classification?

Kuhn suggests that the membership classes of natural categories are altered in the transition between theories.

> The lifelines of [the earth and moon, Mars and Venus] were continuous during the passage from heliocentric to geocentric theory, but the four were differently distributed among natural families as a result of that change. The moon belonged to the family of planets before Copernicus, not afterwards; the earth to the family of planets afterwards, but not before... That sort of redistribution of individuals among natural families or kinds, with its consequent alteration of the features salient to reference, is ... a central ... feature of ... scientific revolutions.[1]

There is a hint of idealism here. For Kuhn asserts that the natural kinds to which the moon and earth belong really did change with change of theory.

Hacking has suggested that such traces of idealism in Kuhn should be taken instead as a form of nominalism.[2] Hacking writes that:

> Kuhn's 'temptation to speak of living in a different world' suggests that he is an idealist, one who holds, in some way, that the mind and its ideas determine the structure of our world. I think he is no idealist, and urge that we should think not of the post-Kantian realist/idealist dichotomy, but of the older, scholastic, realism/nominalism distinction. Kuhn is not among those who challenge the absolute existence of scientific entities or phenomena, nor among those who query the truth conditions for theoretical propositions. Instead he believes that the classifications, categories and possible descriptions that we deploy are very much of our own devising. (1984, pp. 116-7)

For Kuhn, change of taxonomy takes place against the background of general conceptual and categorial stability. Thus Hacking suggests that:

> Kuhn ... might be called an empirical realist and transcendental nominalist. That is, a great many of our commonplace sortings are a given fact of the interactions of any human group and the world in which it lives. That is the empirical realism. But the higher level theories which determine how we think of the world and much of what we do in it and with it are the product of the active power of human minds and collective interaction... On this view, which I call transcendental nominalism, there is not some uniquely right conceptualization of the world, nor is the world of itself constituted by more than merely superficial "kinds of things." The "kinds" that enter our theoretical speculations are man-made... (1979, p. 230)

Hacking's Kuhn is a nominalist about variable theoretical kinds and a realist about stable low-level kinds. It is a "revolutionary nominalism" since "transitions in systems of categories occur during ... revolutionary breaks" (Hacking, 1984, p. 117).

Such nominalism is amenable to two interpretations. The first resembles classical nominalism: i.e. the doctrine that there are no kinds independent of classification. On this interpretation, Kuhn's nominalism consists in the denial that supra-empirical kinds exist independently of scientific classification. The second approach is agnostic about the categorial structure of the world:

the world may have a built-in taxonomy for all we know. But such a taxonomy is epistemically inaccessible to us, so the only kinds which exist as far as science is concerned are those brought into being by a system of classification. On either approach the kinds dealt with by theories are made rather than found.

The difference between the two approaches is ontological versus epistemological: one denies the existence of mind-independent kinds, the other denies knowledge of them. It is unclear which way to interpret Kuhn. Hacking favours the former, for he says: "Kuhn does teach a certain relativism, that there is no uniquely right categorization of any aspect of nature" (1983, p. 110). This way of taking Kuhn can be given textual support. Consider, for example, the following quote from the 'Postscript' to Kuhn's (1970a):

> There is, I think, no theory-independent way to reconstruct phrases like 'really there'; the notion of a match between the ontology of a theory and its "real" counterpart in nature now seems to me illusive in principle. (1970a, p. 206)

On the other hand, in his response to Boyd (1979), Kuhn objects to what he takes to be Boyd's view of science as "zeroing in" on the "world's joints". In that context, Kuhn comments that the 'world with its joints seems to me, like Kant's "things in themselves," in principle unknowable' (1979, p. 418).

Nominalism avoids positing multiple realities, so avoids idealism. It raises the ontological issue of whether higher-level kinds are real or artificial. But the question of the ontological status of such kinds does not, as such, add any new dimension to the problem of incommensurability. Nominalism provides no further reason for taking conceptually disparate theories to be incomparable. For nominalism there is one real world classifiable in diverse ways. To deny the independent existence of classes is not to deny that terms may refer to the same objects. Whether or not classes exist independently, theories may still have intersection of reference with respect to a shared domain of objects. Nor does nominalism introduce any new complexity about translation. The ontological status of higher-order categories is a separate issue from the untranslatability of theoretical concepts. Failure to translate is a relation between systems of concepts, and does not depend upon the way the world is.

Apart from its irrelevance to incommensurability, however, such a restricted nominalism is itself objectionable: for how can there be low-level but not high-level kinds? Let us assume, with

Hacking, that Kuhn is an "empirical realist" about ordinary kinds of things. Let us take the alloys as an invariant lower order kind and the compounds as a higher order variable kind. This is justified on the basis of Kuhn's assumption that the alloys persist as a class while the compounds do not.

If the alloys are a real natural kind, the fact that they constitute a kind depends on facts about their compositional structure. In particular, the fact that a metallic substance is a mixture of metals is what qualifies it for membership in the class of alloys. If what it takes to be a compound is for the molecules of combined elements to enter a chemical bond, then either alloys are compound or they are not. For either their molecules are so bonded or they are not. Consider bronze, for example. It is an objective permanent fact about bronze that its constituent copper and tin molecules do not form a chemical bond. Such facts about the composition of alloys are facts which determine the status of alloys as compounds or mixtures. Because the molecules of alloys are not bonded, as a matter of fact the alloys are not compounds. In this way facts about low-level kinds give rise to certain facts about higher level kinds.

This leads to another problem. On the assumption that there are no facts of the matter about the higher order categories which theories describe, the nominalist could hold that each theory brings its own higher order kinds of things into being. It might then be claimed that theories are about their own unique sets of kinds and that there are no higher order theory-independent kinds for rival theories to disagree about.

But the above argument shows that, if there are low-level kinds, there must be facts about higher order kinds. So, given "empirical realism" about low-level kinds, there do exist higher order kinds about which theories may make conflicting statements.

Moreover, if there are low-level kinds and a shared background language, the joint domain of rival theories can be specified in the background language using neutral terms for low-level kinds such as alloys. But if a joint domain may be independently specified in a background language by reference to the low-level kinds, then by the above argument there exist higher order kinds within the domain to which both theories are relevant. But then there are objective higher order kinds in the domain for conflicting conceptual systems to be about.

It might be objected that, though there are objective facts about the world's categorial structure, to the extent that such facts transcend the empirical level they are unknowable. Theories impose different orderings on the supra-empirical world: since there is no way to check those orderings against real categories, the kinds posited by theories are the only higher order kinds of any significance to us.

What seems right about this objection is that we are unable to step out of our own concepts altogether and directly calibrate them with nature's own categories. And since empirical counterexamples to our theories are typically deflected onto peripheral assumptions the central conceptual apparatus will be protected from direct refutation.

One might argue for more than this by claiming that different theories can only be about the categorial systems which their conceptual apparatus describes. Nature's true categories are beyond our ken and hence irrelevant. There may be no comparison of rival theories because they are about different categorial structures.

But this seems wrong: if theories are about objective structures because of connections with a common domain specified in a background language, then they are about the same objective categorial structure and can be compared relative to it. It is a question of determining which theory-independent kinds the concepts of rival theories are concepts of. Such judgments depend on theories and are fallible. But the fact that mistakes can be made and that theory is needed to specify the kinds does not imply that the judgments cannot be made. Nor does it imply that the only kinds relevant to theory evaluation are the categories which opposing theories themselves posit.

By making use of the background language to specify the joint domain of application, the portion of the world whose categorial structure is in question can be specified. By taking the background language as a metalanguage containing the rival frameworks, particular applications of a concept from one framework may be matched up with the concept applied in the same context by the other. It can in this manner be established that, rather than creating non-overlapping sets of kinds, they apply different concepts to the same set of objects.

In view of these objections the coherence of a restricted nominalism is doubtful. But it is difficult to see why Kuhn should be drawn into nominalism. For nothing follows from the point that theories employ divergent conceptual apparatus about the ontological status of the kinds posited by such apparatus.

There is no reason for Kuhn to take a stand on the ontological issue of the status of kinds at all.

7.3 Conceptual relativism

The problem of multiple systems of kinds may be approached from another direction. Instead of taking it as an ontological issue about the independent existence of kinds, it may be seen as an epistemological issue. That is, the point is not whether kinds are constructed by us, but that judgments about kinds necessarily involve a conceptual perspective.

This point is made in connection with the issue of idealism in the following comment by Rorty:

> the clamor about "idealism" is a red herring. It is one thing to say (absurdly) that we make objects by using words and something quite different to say that we do not know how to find a way of describing an enduring matrix of past and future inquiry into nature except in our own terms — thereby begging the question against "alternative conceptual schemes". Almost no one wishes to say the former. To say the latter is, when disjoined from scary rhetoric about "losing touch with the world," just a way of saying that our present views about nature are our only guide in talking about the relation between nature and our words. (1980, p. 276)

The point may be taken to suggest an argument for conceptual relativism, though it is not so taken by Rorty.[3] For if we have to approach the relation between our words and reality from a conceptual perspective, then that suggests we have no way of breaking out of our present conceptual scheme. That in turn may seem to provide the starting point for an epistemological argument that there is no way to objectively compare or assess different systems of concepts.

Sense perception provides only indirect access to the world, so there is no unmediated access to objects. Thus we have to consider the nature of environing objects always from within some perspective. But the only perspectives available to us are couched in a language with a conceptual apparatus. So we cannot adopt a neutral viewpoint from which to survey the world to determine the appropriateness of this or that way of arranging objects into categories.

Objects can only be dealt with, for cognitive purposes, once they are brought under descriptions. But described objects must be described in terms of some conceptual framework. Evidently, objects under a description cannot function in a neutral manner which would enable us to stand outside of and compare conceptual frameworks. Rather than comparing alternative conceptual frameworks relative to the neutral set of objects they classify, the true situation is one in which the objects themselves can play no direct role.

To employ Putnam's image, we cannot adopt a "God's eye point of view" (e.g. 1981, p. 50) to see what objects our concepts are really about. Since there is no access to neutral objects, conceptual frameworks must be compared by reference to objects under descriptions. But since opposed frameworks disagree over how to describe the objects they classify, there is no way to compare frameworks as classifications of the same objects.

In the final analysis, there are just alternative frameworks with the kinds of things which they divide the world up into, and there is no possibility of an objective comparison of their systems of classification. So we arrive at a conceptual relativism according to which alternative categorial systems cannot be shown to be better or worse conceptualizations because they cannot be compared by appeal to neutral objects.

This argument is flawed in a way which permits defence of the view that frameworks applied in the same domain must have some of the same objects in the extension of some of their terms. The argument depends on the fallacy that the inability to step outside of all conceptual perspectives prevents us from stepping outside of particular conceptual frameworks to take up an external conceptual perspective with respect to them. But that clearly does not follow, for the original point was only that we have inevitably to take up some perspective, not that we are restricted to any particular one.

The argument can be met without assuming direct access to objects. There is no need to suppose that we may step out of our concepts to survey the objects and consider how rival frameworks classify them. Rather, we may make use of some other conceptual perspective which operates neutrally as between the rival frameworks. From this independent point of view we may describe the domain of objects relative to which the rival frameworks disagree.

The point may be made on the basis of Kuhn's own position. For, according to Kuhn, the local changes of central theoretical concepts leave an unaffected background language neutral with respect to such changes. This background language contains vocabulary of varying degrees of theoreticity, so does not constitute an epistemologically pure theory-free medium of expression. Instead, it is sufficiently rich to specify what the rival frameworks are applied to. To the extent that it can identify the objects to which the conflicting vocabularies apply, it can operate as a metalanguage which specifies the common objects over which the rival classifications range. The possibility of taking up an external perspective undermines the conceptual relativist argument without positing either a God's eye viewpoint or a perfect reflection of reality by language.

7.4 Theory-dependence

Boyd (1984) suggests that Kuhn is a "constructivist antirealist" who endorses the thesis that science is so heavily theory-dependent that the world studied by science is a construction. Boyd reconstructs Kuhn's argument for constructivism as follows:

> Roughly, the constructivist antirealist reasons as follows: The actual methodology of science is profoundly theory-dependent. What scientists count as an acceptable theory, what they count as an observation, which experiments they take to be well designed, which measurement procedures they consider legitimate, what problems they seek to solve, and what sorts of evidence they require before accepting a theory — which are all features of scientific methodology — are in practice determined by the theoretical tradition within which scientists work. What sort of world must there be, the constructivist asks, for this sort of theory-dependent methodology to constitute a vehicle for gaining knowledge? The answer, according to the constructivist, is that the world that scientists study, in some robust sense must be defined or constituted by or "constructed" from the theoretical tradition in which the scientific community in question works. If the world that scientists study were not partly constituted by their theoretical tradition, then, so the argument goes, there would be no way of explaining why the theory-dependent methods that scientists use are a way of finding out what is true. (1984, p. 52)

We may set aside the question whether Boyd captures the reasoning behind constructivism and instead extract from the passage an argument that the world of science is a construction. The essential point is that every aspect of scientific practice is theory-dependent. Sense perception is influenced by the theories we accept and the conceptual apparatus we employ. Observations are only made within a background of theories which determine their relevance and significance. Experimental apparatus and measuring instrumentation are manufactured in accordance with and their readings are interpreted by theories. Experimental techniques and standards all depend on theories. Even the facts depend, for their very existence, on the use of a theory and a language relative to which they are facts.

So science deals with facts and observations whose existence is a result of the constructive processes of science itself. The world of science is therefore a world brought into being by scientific activity. Because theories influence what is taken to be empirical fact in different ways, they therefore construct and refer to their own separate worlds.

A strong version of incommensurability derives from this constructivist position. For radically different theories construct and are isolated in their own worlds, which possess no elements in common. So the languages which describe these worlds can have no meaning or reference in common.

There is much to be said in favour of each of the claims of theory-dependence. Trouble begins with taking their cumulative weight to lead to the presumed conclusion. Much of the force of the argument may be removed by insisting pedantically that the consequence of profound theory-dependence is misdescribed as a variance of world or reality. To say that theories create their own worlds is an obfuscation. What varies between theories is the generalized description which each gives of the world. That different means of collecting and interpreting evidence are employed in connection with different theories explains why incompatible theories may each appear to have empirical support. But this only shows that the empirical basis of science contains considerable slack.

The existence of empirical slack enables theories with incompatible portrayals of reality to be successfully applied (to a greater or lesser degree) within the same world. There is no genuine sense in which such theories are actually about different worlds. Though it is undeniable that they yield different world-descriptions, via the mediation of their diverse observational and experimental procedures they are applied to the very same world.

This is more than a verbal point. It removes the need to make sense of the idea of being about a world in such a way that there can be more than one world. The type of incommensurability connected with this form of constructivism is a radical failure of co-reference between theories. If it is not open to the constructivist to speak of reference within the world of a theory, then some other account must be given of lack of co-reference. Let us consider whether the theory-dependence of observation and experimentation entails lack of co-reference.

We may concede that theories in the same domain need not involve the same observations or experiments. Different techniques and instrumentation may be employed. Contact may be made with different kinds of objects via the mediation of different empirical procedures. So there may well be failure of co-reference. Tokens of the same term-type employed within the theories could be applied in separate contexts to different things via their involvement with different experimental operations. Or a term might be introduced in the context of one theory to refer to an object involved in its procedures while no term in the other theory refers to the object.

So variability of procedure is consistent with failure to co-refer. But such variability does not entail failure of co-reference. On the assumption that rival theories investigate a shared domain of empirical phenomena, they may be brought to bear on the same objects, even if here and there different observations are made. There might be different techniques for investigating the properties of a cell, or for measuring the velocity of a particle, and the observations of the cell or particle may vary in significance depending on theoretical context. Still, the very same kind of cell or particle may constitute the object of investigation and reference.

Indeed, theory-ladenness of observation entails nothing at all about reference. To raise the question of reference is to ask whether certain sorts of relation hold between the use of language on the one hand, and extralinguistic entities on the other. But that is a different question from the question of observability. The latter is the question whether an object is so situated relative to human sensory apparatus that it can be detected by the unaided use of sense perception. The class of relations of reference is different from the class of relations of observation. To be sure, reference may depend on observation, as for example in ostension. But one need not observe an object to refer to it. We may refer to what is momentarily absent or concealed. It is possible to refer to what it is impossible for us, given our perceptual capabilities, ever to perceive. It is possible for observers to refer to the same thing while observing different effects. The perceptual experience which two observers have of the same object may be qualitatively very different (e.g. x-ray photographs), though each refers to it. Different instruments may measure the same quantity. And in general nothing can be concluded about reference from the fact that observation fails to be neutral.

Now this may seem objectionable to the constructivist. The constructivist might object that, if observation is theory-laden, there can be no neutral way of deciding questions of reference and co-reference. Since there is no way to step outside of theory in making judgments about reality, each theory must decide upon the nature of its own referents. But this mistakes an epistemic point for an ontological one. Even if there is no theory-neutral means to judge whether rival terms co-refer, the relations of reference may still obtain in virtue of objective relationships between language-user and object of reference. To be sure, our views as to what relations obtain and which entities figure as real referents shall always be affected by our theories. But this forces no concession to constructivism. In fallibilist fashion, it can be granted that such judgments are theory-laden and far from certain, without thereby conceding that reference is somehow theory-relative as well.

In this connection, Devitt has pointed out that all judgments about reference are theory-laden. This point undermines the constructivist argument from theory-ladenness to incommensurability. For if all judgments of reference are theory-laden, then it would seem to follow that all theories are incommensurable. But, as Devitt points out:

> theory comparison must always involve some point of view about the domain in question. But this is just to say that theory comparison is theory-laden, which is true even when the most commensurable theories are compared ... the semantic comparison of theories ... is epistemically like all other attempts to understand the world. (1979, pp. 45-6)

There is, then, no special reason deriving from the theory-ladenness of judgments of reference for thinking that incommensurable theories construct their own completely disjoint real domains of reference. For the same would have also to apply for commensurable, which is to say co-referential, theories.

7.5 Object relativity

Unable to sustain the thesis of radical failure of co-reference by appeal to the theory-dependence of observation, the constructivist must argue for the theory-relativity of domain of reference in some other way. One possibility is to argue that the objects to which theories refer are themselves theory-dependent and that they vary with theory. But can constructivism deny the independence of objects without denying the independence of reality and collapsing into idealism?

Commitment to a mind-independent reality need not be commitment to the existence of mind-independent objects. One might deny that the world itself is divided up into objects. Reality may be fully independent of the mental and yet be amorphous in its own right. Objects may be the contribution of our organizational powers.

But this makes the problem into one of general metaphysics. From the denial of the ultimacy of the objects we interact with, nothing follows about the variability of objects relative to theories. It could be the case that humans interact with objects whose existence is partly due to that interaction. Such objects-for-us may somehow arise out of an underlying reality which differs from the one we experience, either in not dividing into objects, or in dividing into different objects. Or perhaps objects-for-us are Kantian phenomenal presentations, and the noumenal objects-themselves transcend our capacity to know them.

In neither case do the objects studied in science depend for their existence on particular theories. For in each case it is a general fact of human experience that there are objects-for-us which it is the role of science to investigate. If the way we carve the world up into objects is a permanent and general feature of human experience, it does not matter whether our carvings are nature's own. The objects so generated will still be neutral posits as between scientific theories.

To be of relevance here, the constructivist must argue that objects may vary with respect to changes at the theoretical level. Feyerabend discusses the case of the transition from the archaic to the classical world-view, in which he claims there to be an evolution of the concept of an object:

> The concept of an object has changed from the concept of an agregate of equi-important perceptible parts to the concept of an imperceptible essence underlying a multitude of deceptive phenomena. (1975, p. 264)

In brief, this is the contrast between an object as the sum of its observed features and as the real essence behind appearances.

This would only have the consequence that objects are variable if objects depend on concepts. That is, if what objects exist is dependent upon our concept of what it is to be an object, then objects would be relative to conception of object. Reflection upon the problem of the individuation of objects suggests an argument that objects are relative in this way.

The constructivist can point out that our practice of object individuation is insensitive to compositional change. That is, the conditions of identity of objects allow for the loss and gain of component parts. So the constitutive parts of an object do not determine its identity as an object. That suggests that objects are not in themselves objects, but are determined as objects by a convention about identity conditions.

Consider the puzzle of the ship of Theseus. Each plank of which the ship was originally made was replaced as it was rebuilt. The ship remained the same ship but failed to remain the same collection of matter. That we consider the ship a material object has nothing to do with which bits of matter it was built of.

A ship is an artifact, so is literally constructed by us. But natural examples exhibit the same continuity of identity through compositional changes. We consider a river to be the same river even if it does not contain the same water from time to time. A young sapling which grows into an aged oak remains the same tree throughout. Even personal identity is puzzling in this context. In the course of a decade the molecules contained in our body are replaced, so we are not the same stuff as we were.

Such examples show objects not to be determined by composition: what they are made of may change. Unless there are primitive indivisible objects the possibility of such change obtains for any object. Since objects do not maintain constant make-up, their status as objects must be due to a convention to consider them as such. This raises the possibility that objects vary relative to our changing conventions about what to count as an object. There would then be no stable, mind-independent set of objects for all theories to investigate.

This line of reasoning, however, is fallacious. It confuses material constitution with object identity. All that is shown by the changing composition of objects is that objects are not to be strictly identified with their temporal stages, and a fortiori not with their particular composition at a given time.

The question of an object's momentary constitution is not a question about the identity of the object. Rather, it is simply a question about what it is composed of at a particular time. Objects endure through time, and undergo continuous transformations of composition without ceasing to be themselves. Just as I am not the particular collection of molecules that happens to be in my body at a time, neither was Theseus's ship just the set of planks of which it was originally built.

Feyerabend's example is a case of two concepts of objects: one conceives objects in terms of their perceptible parts, the other in terms of their essences. The point that objects are not identical with the particular matter they consist of is a more fundamental point about the nature of objects. For it is fundamental to the nature of objects that they exhibit spatio-temporal continuity and causal regularities. Neither of Feyerabend's examples of a concept of an object deny that objects are perduring physical entities in our environment.

The argument that objects are not identical with constant quantities of matter shows something about the nature of objects. It does not make them relative or theory-dependent. While it is true that objects are picked out relative to our interests, their objectivity consists in the fact that the materials that make them up at any given time and the behaviour which characterizes them are independent of us.

7.6 The rejection of reference

Instead of rejecting objects, the constructivist may choose to reject reference. That is, it may be denied that there is a relation of reference which holds between words and extra-linguistic reality.

Of course, without reference the existence of extralinguistic objects is a moot point. For if there is no reference to independent objects, nothing may be said about them. Without reference to extralinguistic objects, the maximal ontological status attributable to the objects spoken about within language is as linguistic constructs posited for the purposes of thought and talk.

To reject reference is to deny that there are relations of reference which obtain between linguistic expressions and extralinguistic entities.[4] There are two ways to reject reference in this sense. The difference between them is the stance they take to the concept of reference itself. The first way to reject reference is to admit the possibility of reference but to deny that it ever actually obtains. The second way is to reject the concept of extralinguistic reference altogether.

The first option is not a denial of the possibility of reference as such. There is not, according to this first approach, any reason in principle why relations of reference should fail to obtain between language and reality. Rather, as a matter of fact, language is pervaded by empty terms which fail to refer in just the way in which reference to ghosts fails.

What could justify such a thesis of wholesale reference failure? Given the existence of an independent reality (6.3) and objects (7.5), it is implausible to say that reference is a possible relation which simply is never instantiated. Such a thesis would amount to a thorough-going scepticism. For if all of our terms fail to refer, then we are completely mistaken about the world around us and fail to refer to anything that actually exists. Such scepticism faces immense problems in accounting for our apparent success in so many practical activities. Nevertheless, in the conclusion of this section I will briefly consider Putnam's meta-induction on the history of science which seems to give the idea support.

The second option is to dismiss the possibility of extralinguistic reference. The idea is not that reference fails contingently. But rather, as a matter of principle, no relation of reference obtains between language and reality.

The most plausible way to eliminate reference as an extra-linguistic relation is to take a disquotational view of reference. On such a view of reference, talk of reference is to be understood as merely a way of talking about expressions of an object-language within a metalanguage. Rather than interpreting reference as a relation between object-linguistic expressions and extralinguistic objects, the notion of reference is defined in terms of relations within a language. That is, the extension of 'reference' is fully defined for a language by a list of meta-linguistic reference specifications of the form "'t*' refers to t", where the reference of 't*' is specified by its metalinguistic translation 't'.[5]

Such a view of reference is analogous to disquotational accounts of truth which dismiss the notion of truth as correspondence to extralinguistic states of affairs and which reduce truth to a device for semantic ascent.[6] According to such an account, truth is merely disquotational in the sense that to attribute truth to a quotation mark named sentence is equivalent to asserting the sentence. The analogy between disquotational reference and disquotational truth is the following. In the same way that there is no more to truth than the relation between sentences exemplified by a T-sentence, there is nothing to reference beyond associating a term with its metalinguistic translation. Just as disquotational truth is not an extralinguistic correspondence relation, disquotational reference is not an extralinguistic reference relation.

The disquotational approach raises complex issues about truth and reference which are beyond the scope of this book. However, a fundamental difficulty with the rejection of extralinguistic reference is readily apparent. It attaches directly to the rejection of reference in virtue of the relation between reference and truth.

The problem is that without extralinguistic reference a satisfactory distinction between true and false reports of fact cannot be sustained. To say that terms refer in the present strictly disquotational sense is to say that they do not refer to anything extralinguistic. Thus a true and a false statement of fact are unable to differ in the sense that the one contains terms which refer to real things while the false one does not. Nor can it be that the true statement reports a state of affairs which obtains in reality while the false statement reports a state of affairs which does not. If there is no reference to anything existing independently of language, then true report of fact cannot differ from falsity by virtue of reporting anything real.

As against this, it might be objected that a disquotational distinction between 'true' and 'false' does enable true statements to be distinguished from false statements. But this is specious: a merely disquotational distinction between truth and falsity is not an adequate distinction between true and false statement of fact. For the present disquotational account of truth denies reference to anything existing independently of language. Such an analysis of truth is unable to differentiate true from false statement of fact by saying that a true one reports a state of affairs which actually obtains, while a false report does not.

Even though such an account may distinguish truth and falsity formally, no substantive distinction between being true and being false is available without reference to extralinguistic states of affairs.[7]

A more specific difficulty with the rejection of reference in the present context is that it deprives constructivism of an account of what it is for theories to have a common subject matter. If theories refer to nothing extralinguistic, sharing a subject matter cannot consist in being about the same domain of independent objects.

Still, it might seem that the notion of having the same domain can be analyzed in terms of the disquotational notion of reference. In the following, to indicate that the disquotational sense of reference is intended, I will use double quotes and speak of the "reference" of a term. Thus the domain of a theory may be determined by specifying the "referents" over which its language ranges, even though such "referents" are not taken to be extralinguistic. A pair of theories may be about the same domain provided that their languages range over the same "referents" in this disquotational sense.

This is problematic, for how are such "referents" to be individuated? No independent relation of reference to extralinguistic objects can be employed to determine that the "reference" of distinct expressions is the same. But perhaps such "reference" is determined by the descriptive content associated with a term.

The trouble is that alternative theories in the same domain may fail to agree about the properties of the objects within the domain. This would result in difference of the descriptive content associated with the terms defined within such theories. If "reference" is fully determined by descriptive content, theories with non-equivalent descriptive content cannot have the same domain.

This objection may be avoided by reducing the extent to which the identification of a "referent" depends on description. In this case, terms may have identical "reference" even if they are associated with non-equivalent descriptions. This assumes, however, that there is something independent of descriptive content in virtue of which "referents" are identical. To disregard difference of description would not otherwise be justifiable. The only way to determine that non-equivalent descriptions are in fact descriptions of the same thing is for there actually to be something independent of description to which it can be shown that descriptions apply. But this implies the existence of an entity independent of either description to which both descriptions may be applied. And this in turn assumes that such descriptions may refer to extralinguistic entities. But this option is unavailable without extralinguistic reference.

Instead of completely rejecting extralinguistic reference, the constructivist might reject reference to the unobservable while granting reference to observable entities. On such an approach, the subject matter of a theory would then be the set of empirical phenomena which is specified as the theory's domain of application. And theories would have the same subject matter if applied to the same empirical domain.

In effect, this takes the relation of reference to be co-extensive with the observability relation. It enforces a semantic division between terms whose referents are observable and those purporting to refer to the unobservable. In section 7.4 reasons were outlined not to take the relations of reference and observability to be the same relation. Given those reasons, independent argument is needed to show that the relations of reference and observability are co-extensive.

One semantical view which leads to a division between observation terms which refer and theoretical terms which do not is a classical form of verificationism. An account of meaning according to which meaning consists exclusively in empirical verification conditions can support the instrumentalist thesis that theoretical terms are not genuinely referring expressions.

However, the constructivist with which we are concerned takes theoretical language to be meaningful independently of experience, so the instrumentalist alternative is unavailable. If, moreover, the relation of reference is not equivalent to the relation of observability, there could not be any general reason of principle according to which independently meaningful theoretical terms must uniformly fail to refer to anything unobservable.

If reference does not fail as a matter of principle, then perhaps it fails as a matter of fact. This possibility recurs to the first option mentioned at the outset of this section. The alternative to rejecting extralinguistic reference as such is simply to say that the relation contingently fails to obtain.

Taken as a general thesis of universal failure to refer, the thesis is unpromising. Unless it is conjoined with an idealistic rejection of extralinguistic reality, it is implausible to hold that there is universal reference failure for merely epistemic reasons. For that is what it amounts to: if reference is possible but never obtains, that means we are completely mistaken about the world around us.

One argument for such reference failure stems from reflection upon the history of science. It starts from the point that many past theories described entities which do not exist from the perspective of present-day science. Then it is argued that in the same way that past theories have been replaced, present theories also are bound to be rejected. And from the point of view of later theories our own present theories will seem to describe non-existent entities. So not only did older theories fail to refer, even our own theories do not manage to refer. Thus, as Putnam expresses the point:

> eventually the following meta-induction becomes overwhelmingly compelling: just as no term used in the science of more than fifty (or whatever) years ago referred, so it will turn out that no term used now (except maybe observational terms, if there are such) refers. (1978, p. 25)

In the first instance, the meta-induction applies only to theoretical terms. To extend the argument to the observational level there would have to be special reason to claim failure of past reference at the observational level. But the premise of the argument is the point that past theoretical entities failed to exist. This premise does not entail anything about past observable entities.

Nor is there reason to suppose that the premise could be extended to the observational level. What plausibility the premise has with regard to theoretical terms comes from the fact that past theories have indeed posited entities which we now think not to exist. But this is not something we hold to be the case about the ordinary observable entities which our ancestors dealt with in their practical interactions with the world.

Finally, the meta-induction may be criticized even at the theoretical level. For while past and present theories may fail to refer to some theoretical entities, to infer universal error at the theoretical level is to assume a description theory of reference. To say that a past theoretical term was empty because its defining description was false is to assume that misdescription entails reference failure. But, to the contrary, theoretical terms are capable of reference even if their referents are incorrectly described.

Notes

1. Kuhn (1979, p. 417, cf. 1981, p. 25).
2. See Hacking (1979, pp. 229-30, 1983, pp. 109-10 and 1984, pp. 116-7).
3. Rorty appears to endorse Davidson's attack on the scheme-content dualism as an objection to conceptual relativism (1980, pp. 295-305).
4. Thus to reject reference is not to say, with Quine, that reference is inscrutable. The thesis of inscrutability rejects neither reference nor objects: it denies that reference is determinate. Inscrutability is a problem of global dimensions compared with which the problem of co-reference of theories is parochial. The argument for inscrutability depends on the indeterminacy of translating a language's individuative apparatus (see Quine, 1969, pp. 32-5). But theoretical sub-languages are embedded within English which serves as a background language. Thus the individuative apparatus of English is common ground between such theories.
5. For such a definition see Leeds (1978, p. 112), who speaks of "Tarski R sentences" such as "'Caesar' refers to Caesar". In contrast with the view considered here, Leeds' disquotational approach is set within the context of a denial of the determinacy of reference rather than a rejection of reference.
6. On disquotation and semantic ascent see Putnam (1978, pp. 9-10); on disquotational accounts of truth, see Devitt and Sterelny (1987, pp. 162-5).
7. For a parallel objection to the redundancy theory of truth, see Grayling (1982, pp. 156-7), who remarks that interest in the truth of a sentence is motivated by the desire to know if what it says is the case. Cf. Devitt and Sterelny (1987, pp. 168-70) on truth and communication.

Conclusion

In putting forward the thesis of incommensurability Kuhn and Feyerabend draw attention to complex issues concerning the phenomenon of conceptual change in science. Above all, they raise serious problems about the semantical and logical relations between the content of theories which deploy unlike systems of concepts. Yet few of the more extreme claims associated with incommensurability stand scrutiny.

The argument advanced in preceding chapters may be stated in condensed form as follows. I have allowed that the terms of theories with different concepts diverge semantically, and I have argued that the languages of such theories may not be fully intertranslatable. But, given the possibility of referential overlap, it does not follow that the content of such theories is incomparable. Nor, given the distinction between understanding and translation, does it follow that proponents of theories with mutually untranslatable languages are unable to communicate. Neither is it the case that the shift between conceptually divergent theories involves a discontinuous transition between theories which have no common reference or which refer to distinct worlds of their own making. The semantical differences resulting from conceptual variance may be embraced within a thoroughly realist framework on which they are construed as diverse linguistic relations to a fixed and independent reality.

The overall thrust of my argument in this book is deflationary. Incommensurability is less of a problem than has generally been thought. The conceptual and semantical variance which initially gave rise to the idea of incommensurability do not threaten an unmitigated relativism of radically incomparable conceptual schemes. Nor do they force any concession upon an essentially realist view of the relation between scientific theory and extra-theoretic reality.

Consider the claim that incommensurable theories are unable to be compared with one another in respect of content. This is one of the key constitutive claims of the incommensurability thesis. Yet we have found no reason to take incomparability of content as the inevitable result of conceptual disparity between theories. According to the referential approach to content comparison espoused in Chapter Two, various relations of referential overlap may obtain between the terms of such theories which enable their content to be compared. There are, as we saw, severe difficulties with sustaining this referential account of comparison within the framework of a description theory of reference. However, given the modified causal theory of reference adopted here, the claim that major conceptual alterations prevent the content of theories from being compared is without foundation. For, by and large, the transition between conceptually variant theories involves a modicum of referential continuity and overlap. Thus the referential approach removes, or at least seriously undermines, the claim of content incomparability, one of the more extreme claims characteristically associated with the incommensurability thesis.

The principal objective of the last three chapters has been to reinforce the referential approach by showing that the change of reference between incommensurable theories is less extreme than is sometimes suggested. As we saw in Chapter Five, Feyerabend is unable to sustain the thesis of radical reference change, for it cannot be reconciled with the referential continuity at the observational level to which he is committed by the pragmatic theory of observation. And while Kuhn was originally somewhat ambivalent on the nature and extent of referential change, his later account is a restricted thesis of referential variance for higher-level theoretical terms. The criticism of the idealist interpretation in Chapter Six and of constructivism in Chapter Seven further supports the view that purportedly incommensurable theories differ only in having divergent referential relations to a common theory-independent reality: such theories do not refer to distinct theoretically determined "worlds". Taken together, these three chapters provide additional underpinning for the referential approach to content comparison of Chapter Two. For if there is no radical discontinuity of reference and theories do refer to a common world, then sufficient relations of referential overlap may obtain for the purposes of content comparison.

The idea of translation failure between theories is another aspect of the incommensurability thesis which has seemed deeply problematic. This is largely due to its apparent implication of communication failure between rival theorists and their consequent inability to compare their theories. As we saw in Chapter Four, these problems receive trenchant expression in the objections of Putnam and Davidson that the very idea of such translation failure is incoherent. As against such objections, the approach to untranslatability developed in Chapters Three and Four was in part designed to remove the air of paradox which surrounds the idea of untranslatability between theories. In Chapter Three, the positive argument for translation failure was set within the framework of the modified causal theory of reference, which allows referential overlap and comparison even in the absence of translation. In Chapter Four two main points were raised in the course of defending untranslatability against the charge of incoherence: viz. the untranslatability at issue is a restricted translation failure between parts of a language, and understanding a language does not require translation of it. Thus, on the approach to untranslatability developed in these two chapters, rival theorists may communicate and their theories be compared, even if full translation between the special languages of their theories is impossible.

In sum, although the incommensurability thesis raises genuine problems about conceptual change in science, its implications for the philosophy of science are less extreme than has often appeared to be the case. Reference change is not so radical as to preclude referential connections between theories. Translation failure prevents neither communication nor comparison. And the scientific realist may construe semantically variant theories as theories whose terminology refers to a common theory-independent reality in a variety of different ways.

The deflationary import of the approach developed here is further apparent from its implications with regard to the issues of rationality and progress in science. It was noted in the introductory discussion of section 1.1 that both the rationality of theory choice and the progressiveness of theory change are rendered problematic in the light of incommensurability. Given the narrow focus of this book on the problem of incommensurability as such, no attempt has been made to address these issues directly. Yet the problems raised by incommensurability in respect of progress and rationality now appear far less intractable than at first seemed to be the case.

In particular, incommensurability no longer poses a serious threat to rational theory choice given that the content of conceptually disparate theories may be compared and that there may be communication between the advocates of such theories. The various referential overlap relations which may obtain between theories ensure that appropriately related statements from such theories may be compared with respect to agreement and disagreement; so that empirical evidence may support one while disconfirming the other. And since inability to translate from one theory into another does not entail inability to understand a rival theory, it is possible for theoretical adversaries to communicate and engage in rational debate. As for progress, given that theory change involves at least a modicum of referential overlap and continuity, there is no reason to take the transition between conceptually variant theories to be in principle incapable of resulting in progress. Continuity and overlap of reference in the transition between such theories enables later theories to contribute to the stock of truths which has accumulated about some of the same entities that earlier theories referred to.

Where, finally, does this leave the notion of incommensurability? Since so few of the radical claims associated with the incommensurability thesis are warranted by the phenomenon of conceptual change in science, it is not clear that there is anything left for the word 'incommensurability' to stand for. If we like, the word may be retained as a loose name for a cluster of related problems having to do with conceptual change. But there seems little point in saying that theories are incommensurable.

Bibliography

Achinstein, Peter (1964), 'On the Meaning of Scientific Terms', *Journal of Philosophy*, 61, 497-509

Blackburn, Simon (1984), *Spreading the Word*, Oxford University Press, Oxford

Boyd, Richard (1979), 'Metaphor and Theory Change: What is "Metaphor" a Metaphor for?', in A. Ortony (ed.), *Metaphor and Thought*, Cambridge University Press, Cambridge, 356-408

Boyd, R. (1984), 'The Current Status of Scientifc Realism', in J. Leplin (ed.), *Scientific Realism*, University of California Press, Berkeley, 41-82

Brown, Harold I. (1983a), 'Incommensurability', *Inquiry*, 26, 3-29

Brown, H.I. (1983b), 'Response to Siegel', *Synthese*, 56, 91-105

Carnap, Rudolf (1956), *Meaning and Necessity*, The University of Chicago Press, Chicago

Cedarbaum, Daniel G. (1983), 'Paradigms', *Studies in History and Philosophy of Science*, 14, 173-213

Claggett, Marshall (1959), *The Science of Mechanics in the Middle Ages,* The University of Wisconsin Press, Madison

Conant, James Bryant (1964), *The Overthrow of the Phlogiston Theory*, Harvard University Press, Cambridge, Massachusetts

Davidson, Donald (1984), 'On the Very Idea of a Conceptual Scheme', in *Inquiries into Truth and Interpretation*, Clarendon Press, Oxford, 183-198

Devitt, Michael (1979), 'Against Incommensurability', *Australasian Journal of Philosophy*, 57, 29-50
Devitt, M. (1981), *Designation*, Columbia University Press, New York
Devitt, M. (1984), *Realism and Truth*, Blackwell, Oxford
Devitt, M., & Kim Sterelny (1987), *Language and Reality*, Blackwell, Oxford
Dilworth, Craig (1981), *Scientific Progress*, Reidel, Dordrecht
Donnellan, Keith (1977), 'Reference and Definite Descriptions', in S. Schwartz (ed.), *Naming, Necessity and Natural Kinds*, Cornell University Press, Ithaca, 42-65
Dupre, John (1981), 'Natural Kinds and Biological Taxa', *Philosophical Review*, 90, 66-90
Enç, Berent (1976), 'Reference of Theoretical Terms', *Nous*, 10, 261-282
English, Jane (1978), 'Partial Interpretation and Meaning Change', *Journal of Philosophy*, 75, 57-76
Feyerabend, Paul K. (1965), 'Problems of Empiricism', in R.G. Colodny (ed.), *Beyond the Edge of Certainty*, Prentice-Hall, New Jersey, 145-260
Feyerabend, P.K. (1975), *Against Method*, New Left Books, London
Feyerabend, P.K. (1978), *Science in a Free Society*, New Left Books, London
Feyerabend, P.K. (1981a), *Realism, Rationalism and Scientific Method: Philosophical Papers, Vol. 1*, Cambridge University Press, Cambridge
Feyerabend, P.K. (1981b), 'Introduction: Scientific Realism and Philosophical Realism', in (1981a), 3-16
Feyerabend, P.K. (1981c), 'An Attempt at a Realist Interpretation of Experience', in (1981a), 17-36
Feyerabend, P.K. (1981d), 'Explanation, Reduction and Empiricism', in (1981a), 44-96
Feyerabend, P.K. (1981e), "On the 'Meaning' of Scientific Terms", in (1981a), 97-103
Feyerabend, P.K. (1981f), 'Reply to Criticism', in (1981a), 104-131
Feyerabend, P.K. (1981g), *Problems of Empiricism: Philosophical Papers, Vol. 2*, Cambridge University Press, Cambridge
Feyerabend, P.K. (1981h), 'Consolations for the Specialist', in (1981g), 131-167
Feyerabend, P.K. (1987), 'Putnam on Incommensurability', *British Journal for the Philosophy of Science*, 38, 75-81

Field, Hartry (1973), 'Theory Change and the Indeterminacy of Reference', *Journal of Philosophy*, 70, 462-481
Fine, Arthur (1975), 'How to Compare Theories: Reference and Change', *Nous*, 9, 17-32
Fine, A. (1984), 'And Not Anti-Realism Either', *Nous*, 19, 51-65
Grandy, Richard (1983), 'Incommensurability: Kinds and Causes', *Philosophica*, 32, 7-24
Grayling, A.C. (1982), *An Introduction to Philosophical Logic*, Harvester Press, Sussex
Hacking, Ian (1979), 'Review of *The Essential Tension*', *History and Theory*, 18, 223-236
Hacking, I. (1983), *Representing and Intervening*, Cambridge University Press, Cambridge
Hacking, I. (1984), 'Five Parables', in R. Rorty, J.B. Schneewind, & Q. Skinner (eds.), *Philosophy in History*, Cambridge University Press, Cambridge, 103-124
Hankins, Thomas L. (1985), *Science and the Enlightenment*, Cambridge University Press, Cambridge
Hanson, N.R. et al, (1970), 'Discussion at the Conference on Correspondence Rules', in M. Radner & S. Winokur (eds.), *Minnesota Studies in the Philosophy of Science, Vol. IV*, University of Minnesota Press, Minneapolis 220-259
Hesse, Mary B. (1983), "Comment on Kuhn's 'Commensurability, Comparability, Communicability'", in P.D. Asquith & T. Nickles (eds.), *PSA 1982, Vol. 2*, Philosophy of Science Association, East Lansing, Michigan, 704-711
Kitcher, Philip (1978), 'Theories, Theorists and Theoretical Change', *The Philosophical Review*, 87, 519-547
Kitcher, P. (1982), 'Genes', *British Journal for the Philosophy of Science*, 33, 337-359
Kitcher, P. (1983), 'Implications of Incommensurability', in P.D. Asquith & T. Nickles (eds.), *PSA 1982, Vol. 2*, Philosophy of Science Association, East Lansing, Michigan, 689-703
Kordig, Carl R. (1971), *The Justification of Scientific Change*, Reidel, Dordrecht
Kripke, Saul (1972), 'Naming and Necessity' in D. Davidson & G. Harman (eds.) *The Semantics of Natural Languages*, Reidel, Dordrecht
Kroon, Frederick W. (1985), 'Theoretical Terms and the Causal View of Reference', *Australasian Journal of Philosophy*, 63, 143-166
Kuhn, Thomas S. (1970a), *The Structure of Scientific Revolutions*, University of Chicago Press, Chicago, 2nd ed.

Kuhn, T.S. (1970b), 'Reflections on my Critics', in I. Lakatos & A.E. Musgrave (eds.), *Criticism and the Growth of Knowledge*, Cambridge University Press, Cambridge, 231-278

Kuhn, T.S. (1976), 'Theory-Change as Structure-Change: Comments on the Sneed Formalism', *Erkenntnis*, 10, 179-199

Kuhn, T.S. (1977), *The Essential Tension*, University of Chicago Press, Chicago

Kuhn, T.S. (1977a), 'Second Thoughts on Paradigms', in (1977), 293-319

Kuhn, T.S. (1977b), 'Objectivity, Value Judgment, and Theory Choice', in (1977), 320-339

Kuhn, T.S. (1979), 'Metaphor in Science', in A. Ortony (ed.), *Metaphor and Thought*, Cambridge University Press, Cambridge, 409-419

Kuhn, T.S. (1981), 'What Are Scientific Revolutions?', Occasional Paper #18, Center for Cognitive Science, Massachusetts Institute of Technology, Cambridge (Reprinted in *The Probabilistic Revolution, Vol. 1*, L. Krüger, L.J. Daston, & M. Heidelberger, MIT Press, Cambridge, 1987, 7-22)

Kuhn, T.S. (1983), 'Commensurability, Comparability, Communicability', in P.D. Asquith & T. Nickles (eds.), *PSA 1982, Vol. 2*, Philosophy of Science Association, East Lansing, Michigan, 669-688

Laudan, Larry (1976), 'Two Dogmas of Methodology', *Philosophy of Science*, 43, 585-597

Laudan, L. (1977), *Progress and its Problems*, University of California Press, Berkeley

Leeds, Stephen (1978), 'Theories of Reference and Truth', *Erkenntnis*, 13, 111-129

Le Grand, Homer E. (1987), *Revolution #1: Chemistry* (Multilithed source material for History & Philosophy of Science 103, University of Melbourne)

Leplin, Jarrett (1979), 'Reference and Scientific Realism', *Studies in History and Philosophy of Science*, 10, 265-284

Levin, Michael E. (1979), 'On Theory-Change and Meaning-Change', *Philosophy of Science*, 46, 407-424

Mandelbaum, Maurice (1982), Subjective, Objective, and Conceptual Relativisms', in J.W. Meiland & M. Krausz (eds.), *Relativism: Cognitive and Moral*, University of Notre Dame Press, Indiana, 34-61

Martin, Michael (1971), 'Referential Variance and Scientific Objectivity', *British Journal for the Philosophy of Science*, 22, 17-26

Martin, M. (1972), 'Ontological Variance and Scientific Objectivity', *British Journal for the Philosophy of Science*, 23, 252-256

Masterman, Margaret (1970), 'The Nature of a Paradigm', in I. Lakatos & A.E. Musgrave (eds.), *Criticism and the Growth of Knowledge*, Cambridge University Press, Cambridge, 59-89

Mellor, D.H. (1977), 'Natural Kinds', *British Journal for the Philosophy of Science*, 28, 299-312

Moberg, Dale W. (1979), 'Are There Rival, Incommensurable Theories?', *Philosophy of Science*, 45, 244-262

Musgrave, Alan E. (1979), 'How to Avoid Incommensurability', *Acta Philosophica Fennica*, 30, 336-346

Nagel, Ernest (1960), 'The Meaning of Reduction in the Natural Sciences' in A. Danto & S. Morgenbasser (eds.), *Philosophy of Science*, Meridian Books, New York, 288-312

Nagel, E. (1961), *The Structure of Science*, Routledge & Kegan Paul, London

Newton-Smith, W.H. (1981), *The Rationality of Science*, Routledge & Kegan Paul, London

Nola, Robert (1980a), '"Paradigms Lost or the World Regained" — An Excursion into Realism and Idealism in Science', *Synthese*, 45, 317-350

Nola, R. (1980b), 'Fixing the Reference of Theoretical Terms', *Philosophy of Science*, 47, 505-531

Papineau, David (1979), *Theory and Meaning*, Clarendon Press, Oxford

Putnam, Hilary (1962), 'It Ain't Necessarily So', *Journal of Philosophy*, 59, 658-671

Putnam, H. (1975), *Mind, Language and Reality: Philosophical Papers, Vol. 2*, Cambridge University Press, Cambridge

Putnam, H. (1975a), 'Explanation and Reference', in Putnam (1975), 196-214

Putnam, H. (1978), *Meaning and the Moral Sciences*, Routledge & Kegan Paul, London

Putnam, H. (1981), *Reason, Truth and History*, Cambridge University Press, Cambridge

Quine, Willard V.O. (1960), *Word and Object*, MIT Press, Cambridge

Quine, W.V.O. (1969), 'Ontological Relativity', in *Ontological Relativity and Other Essays*, Columbia University Press, New York, 26-68

Rorty, Richard (1980), *Philosophy and the Mirror of Nature*, Blackwell, Oxford

Rorty, R. (1982), 'The World Well Lost', in *Consequences of Pragmatism*, Harvester Press, Sussex, 3-18
Sankey, Howard (1989), *The Incommensurability of Scientific Theories*, PhD thesis, The University of Melbourne
Sankey, H. (1990), 'In Defence of Untranslatability', *Australasian Journal of Philosophy*, 68, 1-21
Sankey, H. (1990-91), 'Feyerabend and the Description Theory of Reference', *Journal of Philosophical Research*, XVI, 223-232
Sankey, H. (1991a), 'Incommensurability and the Indeterminacy of Translation', *Australasian Journal of Philosophy*, 69, 219-223
Sankey, H. (1991b), 'Translation Failure Between Theories', *Studies in History and Philosophy of Science*, 22, 223-236
Sankey, H. (1991c), 'Incommensurability, Translation and Understanding', *The Philosophical Quarterly*, 41, 414-426
Sankey, H. (1992), 'Translation and Languagehood', *Philosophia*, 21, 335-337
Sankey, H. (1993), 'Kuhn's Changing Concept of Incommensurability', *British Journal for the Philosophy of Science*, 44, 775-791
Scheffler, Israel (1967), *Science and Subjectivity*, Bobbs-Merrill, Indianapolis
Shapere, Dudley (1984), *Reason and the Search for Knowledge*, Reidel, Dordrecht
Shapere, D. (1984a), 'The Structure of Scientific Revolutions', in (1984), 37-48
Shapere, D. (1984b), 'Meaning and Scientific Change', in (1984), 58-101
Siegel, Harvey (1987), *Relativism Refuted*, Reidel, Dordrecht
Smith, Peter (1981), *Realism and the Progress of Science*, Cambridge University Press, Cambridge
Sterelny, Kim (1983), 'Natural Kind Terms', *Pacific Philosophical Quarterly*, 64, 110-125
Stroud, Barry (1969), 'Conventionalism and the Indeterminacy of Translation', in D. Davidson & J. Hintikka (eds.), *Words and Objections*, Reidel, Dordrecht, 82-96
Suppe, Frederick (1977), *The Structure of Scientific Theories*, University of Illinois Press, Chicago, 2nd ed.
Zemach, Eddy M. (1976), 'Putnam's Theory on the Reference of Substance Terms', *The Journal of Philosophy*, 73, 116-127